The Open University

ROYAL SOCIETY OF CHEMISTRY

The
Molecular World

Metals and Chemical Change

edited by David Johnson

This publication forms part of an Open University course, S205 *The Molecular World*. Most of the texts which make up this course are shown opposite. Details of this and other Open University courses can be obtained from the Call Centre, PO Box 724, The Open University, Milton Keynes MK7 6ZS, United Kingdom: tel. +44 (0)1908 653231, e-mail ces-gen@open.ac.uk

Alternatively, you may visit the Open University website at http://www.open.ac.uk where you can learn more about the wide range of courses and packs offered at all levels by The Open University.

The Open University, Walton Hall, Milton Keynes, MK7 6AA

First published 2002

Edited, designed and typeset by The Open University.

Published by the Royal Society of Chemistry, Thomas Graham House, Science Park, Milton Road, Cambridge CB4 0WF, UK.

Printed in the United Kingdom by Bath Press Colourbooks, Glasgow.

ISBN 0 85404 665 8

A catalogue record for this book is available from the British Library.

1.1

s205book 4 i1.1

The Molecular World

This series provides a broad foundation in chemistry, introducing its fundamental ideas, principles and techniques, and also demonstrating the central role of chemistry in science and the importance of a molecular approach in biology and the Earth sciences. Each title is attractively presented and illustrated in full colour.

The Molecular World aims to develop an integrated approach, with major themes and concepts in organic, inorganic and physical chemistry, set in the context of chemistry as a whole. The examples given illustrate both the application of chemistry in the natural world and its importance in industry. Case studies, written by acknowledged experts in the field, are used to show how chemistry impinges on topics of social and scientific interest, such as polymers, batteries, catalysis, liquid crystals and forensic science. Interactive multimedia CD-ROMs are included throughout, covering a range of topics such as molecular structures, reaction sequences, spectra and molecular modelling. Electronic questions facilitating revision/consolidation are also used.

The series has been devised as the course material for the Open University Course S205 *The Molecular World*. Details of this and other Open University courses can be obtained from the Course Information and Advice Centre, PO Box 724, The Open University, Milton Keynes MK7 6ZS, UK; Tel +44 (0)1908 653231; e-mail: ces-gen@open.ac.uk. Alternatively, the website at www.open.ac.uk gives more information about the wide range of courses and packs offered at all levels by The Open University.

Further information about this series is available at www.rsc.org/molecularworld.

Orders and enquiries should be sent to:

Sales and Customer Care Department, Royal Society of Chemistry, Thomas Graham House, Science Park, Milton Road, Cambridge, CB4 0WF, UK

Tel: +44 (0)1223 432360; Fax: +44 (0)1223 426017; e-mail: sales@rsc.org

The titles in *The Molecular World* series are:

THE THIRD DIMENSION
   edited by Lesley Smart and Michael Gagan

METALS AND CHEMICAL CHANGE
   edited by David Johnson

CHEMICAL KINETICS AND MECHANISM
   edited by Michael Mortimer and Peter Taylor

MOLECULAR MODELLING AND BONDING
   edited by Elaine Moore

ALKENES AND AROMATICS
   edited by Peter Taylor and Michael Gagan

SEPARATION, PURIFICATION AND IDENTIFICATION
   edited by Lesley Smart

ELEMENTS OF THE p BLOCK
   edited by Charles Harding, David Johnson and Rob Janes

MECHANISM AND SYNTHESIS
   edited by Peter Taylor

The Molecular World Course Team

Course Team Chair
Lesley Smart

Open University Authors
Eleanor Crabb (Book 8)
Michael Gagan (Book 3 and Book 7)
Charles Harding (Book 9)
Rob Janes (Book 9)
David Johnson (Book 2, Book 4 and Book 9)
Elaine Moore (Book 6)
Michael Mortimer (Book 5)
Lesley Smart (Book 1, Book 3 and Book 8)
Peter Taylor (Book 5, Book 7 and Book 10)
Judy Thomas (*Study File*)
Ruth Williams (skills, assessment questions)
Other authors whose previous contributions to the earlier courses S246 and S247 have been invaluable in the preparation of this course: Tim Allott, Alan Bassindale, Stuart Bennett, Keith Bolton, John Coyle, John Emsley, Jim Iley, Ray Jones, Joan Mason, Peter Morrod, Jane Nelson, Malcolm Rose, Richard Taylor, Kiki Warr.

Course Manager
Mike Bullivant

Course Team Assistant
Debbie Gingell

Course Editors
Ian Nuttall
Bina Sharma
Dick Sharp
Peter Twomey

CD-ROM Production
Andrew Bertie
Greg Black
Matthew Brown
Philip Butcher
Chris Denham
Spencer Harben
Peter Mitton
David Palmer

BBC
Rosalind Bain
Stephen Haggard
Melanie Heath
Darren Wycherley
Tim Martin
Jessica Barrington

Course Reader
Cliff Ludman

Course Assessor
Professor Eddie Abel, University of Exeter

Audio and Audiovisual recording
Kirsten Hintner
Andrew Rix

Design
Steve Best
Carl Gibbard
Sarah Hack
Mike Levers
Sian Lewis
John Taylor
Howie Twiner

Library
Judy Thomas

Picture Researchers
Lydia Eaton
Deana Plummer

Technical Assistance
Brandon Cook
Pravin Patel

Consultant Authors
Ronald Dell (*Case Study:* Batteries and Fuel Cells)
Adrian Dobbs (Book 8 and Book 10)
Chris Falshaw (Book 10)
Andrew Galwey (*Case Study:* Acid Rain)
Guy Grant (*Case Study:* Molecular Modelling)
Alan Heaton (*Case Study:* Industrial Organic Chemistry, *Case Study:* Industrial Inorganic Chemistry)
Bob Hill (*Case Study:* Polymers and Gels)
Roger Hill (Book 10)
Anya Hunt (*Case Study:* Forensic Science)
Corrie Imrie (*Case Study:* Liquid Crystals)
Clive McKee (Book 5)
Bob Murray (*Study File*, Book 11)
Andrew Platt (*Case Study:* Forensic Science)
Ray Wallace (*Study File*, Book 11)
Craig Williams (*Case Study:* Zeolites)

CONTENTS

METALS AND CHEMICAL CHANGE

David Johnson and Kiki Warr

CASE STUDY: BATTERIES AND FUEL CELLS

Ronald Dell and David Johnson

Metals and Chemical Change

David Johnson and Kiki Warr

INTRODUCTION

1

In the earlier books in this series, there has been an emphasis on molecular and electronic structure — that is, on the spatial arrangement of atoms within chemical substances, and on the arrangement of electrons within atoms. Very little has been said about chemical change. But here this emphasis shifts, and we ask why chemical reactions happen. There are two conditions that must be fulfilled before a chemical reaction can occur: the equilibrium constant must be sufficiently favourable, and the rate must be sufficiently fast. This Book will be concerned with both conditions, but mainly with the first. You will meet new 'labour-saving' properties of chemical substances; these will allow us to predict whether a chemical reaction has a favourable equilibrium constant or not; the reaction does not even have to be tried out.

The units of the properties in question are mainly those of energy, and come from a branch of science called *thermodynamics*. To show the relevance of this subject, we shall use it to explore an important problem about the chemical behaviour of metals. Finally, when our study of this problem is complete, thermodynamics is used again, towards the end of the Book, in a systematic study of the chemistry of the alkali and alkaline earth elements — that is, of Groups I and II of the Periodic Table. Along with thermodynamics, metals are therefore a major theme in this Book, so we begin by reminding you about them, and about the way that their properties are explained by the simplest theory of metallic bonding.

1.1 Metals and their physical properties

Figure 1.1 shows a full Periodic Table, colour-coded to reveal the periodic distribution of metals, semi-metals and non-metals. Of the 114 known elements, 90 are, or are likely to be, metallic. This, and the other books in the series, concentrate on the 46 typical elements. Here, metals are not so predominant, but, even so, they still outnumber each of the other two categories.

Some metallic elements, such as bismuth and manganese, are brittle, but most, when pure, are **malleable** and **ductile**. Malleable materials are those that can be reshaped by hammering; ductile materials can be drawn out under tension into wires. Figure 1.2 shows a piece of early British gold jewellery dated 1600 BC. Such things were made by hammering out gold into sheets. This is possible because the metal is malleable, and at the same time strong.

Malleability and ductility are especially associated with those metals that possess one of the two close-packed structures discussed in *The Third Dimension: Crystals*[1] *.

● What are the names of these two types?

● Hexagonal close-packed and cubic close-packed.

* See the references in Further Reading (p. 234) for details of other titles in *The Molecular World* series that are relevant.

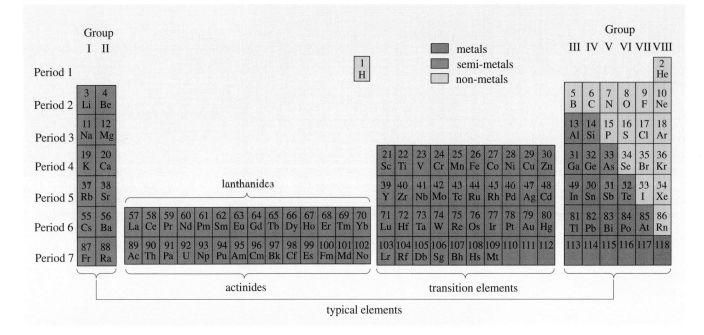

Group
I II

metals
semi-metals
non-metals

Group
III IV V VI VII VIII

Figure 1.1
Long form of the Periodic Table, showing metals, semi-metals and non-metals. Isotopes of elements 110–112, 114 and 116 have been made, but these elements have so far [2002] not been given names.

Figure 1.2
Gold lanula from Gwynedd, ca. 1600 BC.

Both close-packed structures consist of layers of atoms of the metallic element. It follows that if we can explain why such layers can slip over one another fairly easily, we can account for both malleability and ductility. Now a simple model of a metal consists of a regular array of positively charged ions in a 'pool' of freely moving electrons. The interaction between the positive ions and the negatively charged electrons, which surround the ions, holds the metal together. Let us contrast the situations in a metal and in an ionic solid, such as NaCl, when the layers are displaced relative to one another. Look first at Figure 1.3b.

🔘 Why should such a displacement be unfavourable in an ionic solid?

🔘 The displacement brings like charges in adjacent layers into close proximity. Repulsion between the charges will then push the layers apart.

This explains why fracture and not deformation is usually the result of beating an ionic solid. The contrast with the situation in a metal (Figure 1.3a) is obvious: all the ions are of like charge, the situation after displacement is similar to what it was before, and the freely moving electrons can adjust to the change without further disruption. Consequently, in a pure metallic crystal, layers can usually slip easily over one another, thus accounting for the properties of ductility and malleability.

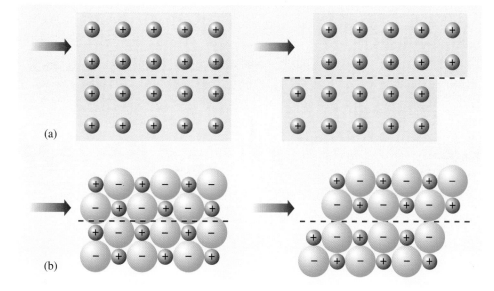

(a)

(b)

Figure 1.3
Two different types of solid undergoing a shearing movement along a plane (red broken line) through the crystal: (a) the crystal of a metal; (b) the crystal of an ionic solid containing positive and negative ions.

You will be familiar with other properties of metals from your everyday contact with iron, aluminium, copper, silver and tin, for example. They often have a lustrous appearance, and are good conductors of heat. However, the most characteristic property of metals is their high electrical conductivity. This is explained by the free electrons that roam at random through the metallic structure. When a voltage difference is applied across two points on a piece of metal, the motion of the electrons becomes less random, there is an overall movement of electrons between the two points, and an electric current flows.

The unit of electrical conductivity is siemens per metre $(S\,m^{-1})$ *. Those elements classified as metals in Figure 1.1 have an electrical conductivity at or below room temperature of at least $3 \times 10^5\,S\,m^{-1}$ along any direction in a single crystal of any known form of the element. Although it is a good electrical conductor, carbon in the

* The electrical conductivity in siemens per metre tells you the expected current in amps when a potential difference of one volt is placed across two opposite faces of a cube of the material with sides of one metre.

form of graphite does not meet this criterion. The structure of graphite is shown in Figure 1.4. It consists of sheets of carbon atoms, and each atom is bonded to three others in the same sheet. Carbon has four outer electrons in the shell structure (2, 4); in graphite, each carbon atom shares three of these with three other carbon atoms in the same sheet by forming three covalent C−C bonds. The fourth electron is mobile, just as the bonding electrons in a metal are mobile; it binds carbon atoms within its sheet more strongly together by contributing to a pool of electrons concentrated around the plane of the sheet. Consequently, there is high conductivity parallel to the sheets, but very low conductivity at right-angles to them.

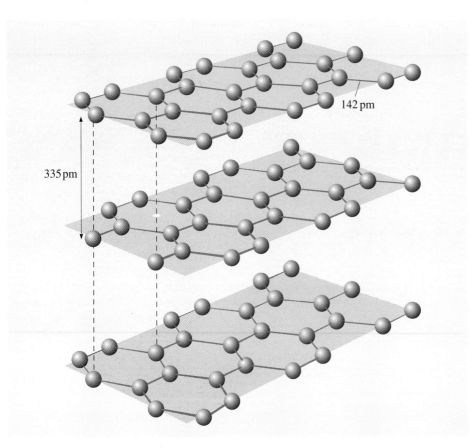

Figure 1.4
The structure of graphite; there is metallic conductivity in directions parallel to the sheets of carbon atoms, but not at right-angles to them.

Because of this property, graphite is sometimes called a *two-dimensional metal*. However, our decision to treat as metals only those elements with high electrical conductivity in all three dimensions rules graphite out, and carbon is not classified as a metal in Figure 1.1.

1.2 Summary of Section 1

1   Most of the chemical elements are metals, and even among the typical elements, metals predominate.

2   Most, but not all, metallic elements are malleable and ductile. Elements classified as metals have high electrical conductivities in all directions in a single crystal of the element. Using this criterion, carbon is a non-metal.

3   These properties can be explained by a model in which metals are regarded as positive ions immersed in a pool of free electrons.

REACTIONS OF METALS

2

Between now and the end of this Book, we shall investigate an important problem about the chemical behaviour of metals. Certain features of the problem are familiar to everybody. Gold jewellery (Figure 2.1) can survive essentially unchanged for thousands of years. Many bronze busts and statues (Figure 2.2) are much more recent, but the green stains on their stone bases show that significant corrosion of their copper content has already taken place. Again, the uranium metal intended for Nazi Germany's first nuclear reactor went up in flames when a physicist took a shovel to it; rubidium reacts violently with water, and inflames in air, without the assistance of a shovel (Figure 2.3). Few chemists would quarrel with you if you said that copper was more reactive than gold, and that rubidium was more reactive than uranium, but what exactly do we mean by the word 'reactive'? Again, is our statement about reactivities true only in moist air, or does it remain correct in the presence of other chemicals? It is these and other questions that we shall examine for the remainder of this Book.

To begin with, you must learn a little more about the reactions of metals. The next three Sections give you the opportunity to do this, partly by reading, and partly by watching experiments on the CD-ROM application, *Reactions of metals*.

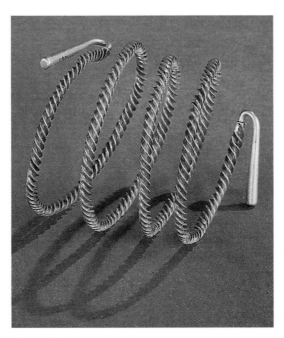

Figure 2.1
Gold torque from Stanton, Staffordshire, ca. 1000 BC.

Figure 2.2
Bronze is about 90% copper and 10% tin; the green stains on the base of this statue of Scotland's hero, William Wallace, are an indication of copper corrosion.

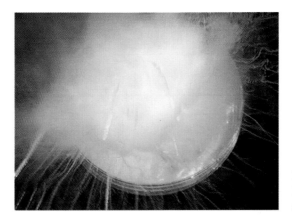

Figure 2.3
Rubidium reacting with water and air.

ACTIVITY 2.1 Introduction: reactions with dilute acid

This activity introduces the problem of metal reactivity, and then asks you to record what happens when the metals copper, iron, magnesium, tin and zinc are added to dilute acid. It is the first sequence of *Reactions of metals* on the CD-ROM associated with this Book.

As noted in the latter part of Activity 2.1, in those cases where a reaction happened, the metal dissolved in the dilute acid to form a dipositive aqueous ion. For example, in the case of iron, this ion is $Fe^{2+}(aq)$. Moreover, when a reaction occurred, bubbles of gas were evolved. This happened for iron, magnesium, tin and zinc. You may remember that magnesium reacts with acids to yield hydrogen gas. Thus, for iron, a likely equation for the reaction is:

$$Fe(s) + 2H^+(aq) = Fe^{2+}(aq) + H_2(g) \tag{2.1}$$

Notice that this equation has been pared down to just the chemical species that *change* during the reaction; species like the chloride ion, $Cl^-(aq)$, which are present but do not change, have been left out.

● Try to write equations for the other three reactions that you observed.

● Following the example of iron in Equation 2.1, you should get:

$$Mg(s) + 2H^+(aq) = Mg^{2+}(aq) + H_2(g) \tag{2.2}$$
$$Sn(s) + 2H^+(aq) = Sn^{2+}(aq) + H_2(g) \tag{2.3}$$
$$Zn(s) + 2H^+(aq) = Zn^{2+}(aq) + H_2(g) \tag{2.4}$$

In Reactions 2.1–2.4, you can see one of the most prominent chemical characteristics of metals.

● What is it?

● They form positive rather than negative ions in aqueous solution.

Metals also tend to form positive ions in solid compounds. For example, we say that solid sodium chloride contains Na^+ ions, and this contrasts with the behaviour of non-metals, which do not form positive ions, and indeed often form monatomic negative ions such as Cl^- or O^{2-}. There are important terms for describing changes of this type, to which we now turn.

2.1 Oxidation and reduction

The words *oxidation* and *reduction* tend to take on broader meaning as one learns more chemistry. You are now ready to take a step down this road. You know that a substance is said to be oxidized when it reacts with oxygen. Thus, magnesium is oxidized when it burns in oxygen gas (Figure 2.4):

$$Mg(s) + \frac{1}{2}O_2(g) = MgO(s) \tag{2.5}$$

Let us consider how electrons are redistributed during this reaction by using an ionic bonding model.

Figure 2.4 Magnesium metal burning in the oxygen of the air.

How are magnesium and oxygen held together in MgO?

Each magnesium atom loses two electrons and forms the ion Mg^{2+}, which has the electronic structure of neon; each oxygen atom gains two electrons and forms the ion O^{2-}, which also has the electronic structure of neon. The forces between the oppositely charged ions in $Mg^{2+}O^{2-}$ hold the compound together.

Thus, when the magnesium atoms on the left of Equation 2.5 are oxidized, they lose electrons and become Mg^{2+} ions in MgO. Chemists fasten on to this electronic change, and use it in a broader definition of **oxidation**:

> The loss of electrons from an atom, compound or ion is described as oxidation.

Another feature of Equation 2.5 is that the oxygen atoms, bound together on the left-hand side as diatomic molecules, take on electrons and form O^{2-} ions. Thus, whereas the magnesium atoms lose electrons, the oxygen atoms gain them. The taking-up of electrons, a process that is the reverse of oxidation, is called **reduction**:

> The gain of electrons by an atom, compound or ion is described as reduction.

A useful mnemonic that will help you to remember this is OILRIG: Oxidation Is Loss; Reduction Is Gain.

In Equation 2.5, notice that oxidation and reduction occur together; they are complementary processes: the magnesium is oxidized and the oxygen is reduced. Apart from certain exceptional cases, a reaction that includes oxidation, also includes reduction, and vice versa. To mark this fact, such reactions are often called **redox reactions**, and Reactions 2.1–2.4 are typical examples.

● In Equations 2.1–2.4, is the metal oxidized or reduced?

● The metal loses electrons and forms positive ions; it has been oxidized.

● In Equations 2.1–2.4, what happens to the hydrogen ions?

● They are reduced; they gain electrons and form neutral hydrogen atoms, which are combined in diatomic molecules.

QUESTION 2.1

Which of the following are redox reactions? In each redox reaction, identify the element that is oxidized, and the element that is reduced:

(i) $K(s) + H^+(aq) = K^+(aq) + \frac{1}{2} H_2(g)$

(ii) $Cu^{2+}(aq) + Fe(s) = Cu(s) + Fe^{2+}(aq)$

(iii) $Mg(s) + F_2(g) = MgF_2(s)$

(iv) $Ca^{2+}(aq) + 2F^-(aq) = CaF_2(s)$

(v) $Cl_2(g) + 2Fe^{2+}(aq) = 2Cl^-(aq) + 2Fe^{3+}(aq)$

STUDY NOTE

A set of interactive self-assessment questions are provided on the *Metals and Chemical Change* CD-ROM. The questions are scored, and you can come back to the questions as many or as few times as you wish in order to improve your score on some or all of them. This is a good way of reinforcing the knowledge you have gained while studying this Book.

2.2 Oxidation of metals by aqueous hydrogen ions

We have now established that Reactions 2.1–2.4 are all redox reactions. Furthermore, in all of them, a metal atom reacts with two hydrogen ions to give an ion with two positive charges and a molecule of hydrogen gas. However, Activity 2.1 revealed a significant difference in the vigour of the reactions. In particular, the magnesium reaction was quite violent; by contrast, the iron, tin and zinc reactions were slow. Indeed, the very slow tin reaction only became perceptible on heating. Similar reactions, which are as violent or more violent than the magnesium reaction, are observed when the alkali metals (Group I of the Periodic Table) react with acids. For example, steady evolution of hydrogen occurs from the surface of lithium metal, but with sodium the gas evolution is extremely vigorous; for potassium, rubidium and caesium, the reaction is explosive, and the hydrogen gas catches fire. The reactions with plain water are very similar.

● Write equations for the reactions of lithium and caesium with acids, in which hydrogen gas is produced.

● Like magnesium, the alkali metals form ions with noble gas structures, losing one electron to form singly charged cations:

$$Li(s) + H^+(aq) = Li^+(aq) + \frac{1}{2}H_2(g) \qquad (2.6)$$

$$Cs(s) + H^+(aq) = Cs^+(aq) + \frac{1}{2}H_2(g) \qquad (2.7)$$

So experiments show that magnesium and the alkali metals react much more violently with aqueous hydrogen ions than do zinc, iron and tin. However, in Activity 2.1 there was one element that did not react at all with dilute hydrochloric acid.

● Which element was this?

● It was copper; when copper is dropped into dilute hydrochloric acid, no hydrogen gas is evolved, and the dipositive aqueous ion, $Cu^{2+}(aq)$, is not formed.

Now hydrated forms of copper sulfate, $CuSO_4$, are used by gardeners as a fungicide. The solution of this compound in water is blue because of the blue ion $Cu^{2+}(aq)$ (Figure 2.5):

$$CuSO_4(s) = Cu^{2+}(aq) + SO_4^{2-}(aq) \qquad (2.8)$$

Thus, the ion $Cu^{2+}(aq)$ exists.

Figure 2.5
Dried copper sulfate, $CuSO_4$ (left) is colourless, but it dissolves in water to form the blue aqueous ion, $Cu^{2+}(aq)$ (centre). On standing in moist air, dried $CuSO_4$ absorbs water and turns blue, forming the solid copper sulphate used by gardeners (right). This has the empirical formula $CuSO_4.5H_2O$.

● Write an equation for a reaction of metallic copper with acid to produce hydrogen gas.

● $$Cu(s) + 2H^+(aq) = Cu^{2+}(aq) + H_2(g) \qquad (2.9)$$

As Activity 2.1 has demonstrated, this reaction does not happen. This is true of the analogous reaction for certain other metals, in particular silver and gold. Silver forms a well-defined colourless ion $Ag^+(aq)$ and, for our purposes, the most convenient description of aqueous oxidized gold is the ion $Au^{3+}(aq)$. Thus, we can write two further equations for reactions that do *not* occur.

● Write them down now!

● $Ag(s) + H^+(aq) = Ag^+(aq) + \frac{1}{2}H_2(g)$ (2.10)

$Au(s) + 3H^+(aq) = Au^{3+}(aq) + \frac{3}{2}H_2(g)$ (2.11)

We now have three metals, copper, silver and gold, which we know can form ions under some circumstances, but cannot do so by a reaction with hydrogen ions. Hydrogen ions can oxidize the alkali metals, magnesium, zinc, iron and tin, but

not copper, silver and gold.

Let us now summarize the observations made in this Section. We can do it by making what seems a somewhat arbitrary classification of the metals that we have discussed, but the classification will later turn out to be useful.

In the first class are those metals that are not oxidized by aqueous hydrogen ions at all. Important examples are copper, silver and gold.

In the second class we shall put metals, such as zinc, which react mildly with hydrogen ions. Iron, which forms a dipositive ion $Fe^{2+}(aq)$ when it reacts with dilute acid, is also a member of this class.

Finally, we are left with a third class, the alkali metals and alkaline earth metals (magnesium, calcium, strontium and barium), which react violently with hydrogen ions. Examples of the behaviour of the three classes with hydrogen ions are illustrated in Figure 2.6.

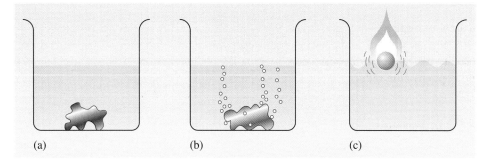

(a)   (b)   (c)

Figure 2.6 Differences in the vigour of the reaction of metals with hydrogen ions cxcmplificd by (a) coppcr, (b) zinc, and (c) potassium.

Now, all the metals that we have considered have been subjected to the same test: an attempt to oxidize them with aqueous hydrogen ions.

● Suggest a grading of the three classes in order of the tendency of the metals to be oxidized to aqueous ions by $H^+(aq)$.

● A reasonable hypothesis is as follows: those metals that are unaffected by dilute acids (the first class) are the least readily oxidized. If we use the violence of the reaction in the two remaining classes as an index of the tendency of the metal to be oxidized to its aqueous ions, then the third class is more readily oxidized than the second. Thus, the grading is third class > second class > first class.

2.3 Reactions of metals with aqueous metal ions

Before reading further, do Activity 2.2.

ACTIVITY 2.2   Reactions with aqueous ions 1

In this activity, the second sequence of *Reactions of metals* on the CD-ROM, you will see the results of adding zinc to a solution of copper sulfate, and copper to a solution of zinc sulfate. Write out an equation for any reaction that occurs.

The reddish-brown film formed on the zinc is copper metal, and the simultaneous fading of the blue colour from the solution is consistent with this, because the metal is formed from the $Cu^{2+}(aq)$ ions in the solution. Because zinc forms colourless $Zn^{2+}(aq)$ ions (see Activity 2.1), a reasonable equation is

$$Zn(s) + Cu^{2+}(aq) = Zn^{2+}(aq) + Cu(s) \qquad (2.12)$$

● What is oxidized and what is reduced during this reaction?

● Zinc atoms lose electrons, whereas copper ions gain them. Thus, zinc metal is oxidized and $Cu^{2+}(aq)$ ions are reduced.

● What happens when copper metal is dropped into a solution of zinc sulfate?

● Nothing: in other words, the reverse of the reaction in Equation 2.12 does not occur.

Copper metal and zinc sulfate solution are the products left when the reaction of zinc metal with copper sulfate solution has ended. The equilibrium position in the system represented by Equation 2.12 lies very far to the right: the favoured combination is zinc in the oxidized form, $Zn^{2+}(aq)$, and copper in the reduced form, $Cu(s)$ (Figure 2.7). This is the combination produced when zinc metal is oxidized at the expense of the 'de-oxidation' (reduction) of $Cu^{2+}(aq)$. So of the two metals — zinc and copper — zinc is the more readily oxidized to aqueous ions. Now this is consistent with Section 2.2, where we concluded that as $H^+(aq)$ will oxidize zinc but not copper, zinc is the more readily oxidized of the two metals. Let us therefore assume that this consistency with the classification of Section 2.2 is of general application.

Figure 2.7 Reaction 2.12 shows that the combination of copper metal and a solution of zinc sulfate (right) is preferred to, or more stable than, the combination of zinc metal and a solution of copper sulfate (left).

In the light of this assumption, predict answers to the following questions. When you have made your predictions, you should do Activity 2.3.

⬤ What do you expect to happen when (i) magnesium is added to a solution containing $Ag^+(aq)$, (ii) copper is added to a solution containing $Mg^{2+}(aq)$, and (iii) magnesium is added to a solution containing $Cu^{2+}(aq)$?

Mg will oxidize, Ag will reduce
nothing will happen
Mg will oxidize, Cu will reduce

ACTIVITY 2.3   Reactions with aqueous ions 2

In this activity, the third sequence of *Reactions with metals*, on the CD-ROM associated with this Book, you will test the predictions that you have just made by watching the three experiments mentioned above.

We now return to the predictions that you were asked to make before you did Activity 2.3. According to the classification of Section 2.2, magnesium is a member of the third class, and silver is a member of the first, so magnesium should be more readily oxidized than silver. The magnesium should be oxidized to $Mg^{2+}(aq)$, and $Ag^+(aq)$ should be reduced to silver metal when magnesium is added to silver nitrate solution.

⬤ Write an equation for the reaction.

⬤ As the ions are of different charge, conservation of charge can be achieved only by making *two* silver ions react with each magnesium atom:

$$Mg(s) + 2Ag^+(aq) = Mg^{2+}(aq) + 2Ag(s) \tag{2.13}$$

Likewise, because copper is in the first class along with silver, magnesium should also reduce $Cu^{2+}(aq)$ to metallic copper:

$$Mg(s) + Cu^{2+}(aq) = Mg^{2+}(aq) + Cu(s) \tag{2.14}$$

This reaction does indeed occur, but the reverse reaction does not: copper undergoes no reaction when dropped into a solution of magnesium sulfate.

We see then that the results of Activities 2.2 and 2.3 are in agreement with the classification made in Section 2.2. Moreover, they suggest an enlarged scope for the classification. Not only do the classes give us an idea of the tendency of a metal to be oxidized by $H^+(aq)$, but there seems a possibility that they may also tell us whether or not a metal will reduce the aqueous ions of a metal in a different class.

Now try Questions 2.2 and 2.3.

QUESTION 2.2

Describe what will happen, if anything, when pieces of the metals rubidium, strontium and silver are dropped into dilute hydrochloric acid. Write equations for any reaction that you think will occur.

QUESTION 2.3

Predict the consequences, if any, of mixing the following chemicals; write equations involving aqueous ions where you think any reaction occurs:

(i)   copper and an aqueous solution of tin dichloride ($SnCl_2$);

(ii) tin and an aqueous solution of silver nitrate;

(iii) magnesium and an aqueous solution of tin dichloride;

(iv) iron and an aqueous solution of calcium chloride.

2.4 Reactions of metals with oxygen and the halogens

In this Section, we use an ionic model in which the solid halides and oxides of metallic elements are regarded as aggregates of positive metal ions and halide (X^-) or oxide (O^{2-}) ions. Consider the formation of two chlorides of this type from the metal and chlorine. Caesium, an alkali metal, forms a chloride CsCl; gold, a noble metal, forms a chloride, $AuCl_3$:

$$Cs(s) + \tfrac{1}{2}Cl_2(g) = CsCl(s) \tag{2.15}$$

$$Au(s) + \tfrac{3}{2}Cl_2(g) = AuCl_3(s) \tag{2.16}$$

In the ionic model, CsCl is regarded as Cs^+Cl^-, and $AuCl_3$ as $Au^{3+}(Cl^-)_3$.

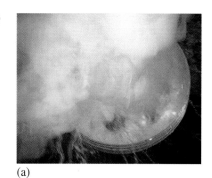
(a)

What happens to the metal and chlorine in Reactions 2.15 and 2.16?

Several things, but the relevant one here is that the metal is oxidized and the chlorine is reduced. The metal atoms lose electrons and become positive ions in the chloride; the chlorine atoms, bound together in diatomic molecules, gain electrons and become Cl^- ions. As usual, oxidation and reduction are complementary processes.

Although both caesium and gold react with chlorine gas, there is an enormous difference in the vigour of the two reactions. At room temperature, the reaction of gold is very slow, but caesium explodes when exposed to a plentiful supply of chlorine. When the two chlorides are gently heated, nothing happens to caesium chloride, but the gold compound decomposes, and the metal and chlorine gas are regenerated:

$$AuCl_3(s) = Au(s) + \tfrac{3}{2}Cl_2(g) \tag{2.17}$$

These observations suggest that gold has a much lesser tendency than caesium to be oxidized by chlorine; not only does it combine with the halogen less vigorously, but the metal can easily be regenerated from the chloride.

A very similar pattern is observed in the reactions of the two metals with the other halogens, and with oxygen. Caesium reacts explosively with fluorine and oxygen, and combines violently with bromine and iodine when gently heated. By contrast, gold can be melted in air or oxygen without chemical change (Figure 2.8), and combines with bromine and iodine in mild reactions to give halides that, like the trichloride, regenerate the metal on heating. In addition, we know that caesium reacts explosively with dilute acids and water; gold, on the other hand, is unaffected.

A chemical that brings about oxidation is called an **oxidizing agent**; one that brings about reduction is called a **reducing agent**.

We have now considered seven oxidizing agents — the four halogens, oxygen, water and the aqueous hydrogen ion. Whichever oxidizing agent we choose, the reactivity of caesium metal is always greater than that of gold. Suppose now that this result could be generalized to include all metals. It would then be possible to produce a *unique* grading of metals in order of their tendency to be oxidized, a grading that is valid whether the oxidizing agent is aqueous hydrogen ion, chlorine, bromine, oxygen, or whatever. If this were the case, then the classification that we made in Section 2.2 in terms of the tendency to be oxidized by hydrogen ions would also

(b)

Figure 2.8
(a) Caesium reacts instantly with water and air; (b) gold can be melted at over 1 000 °C in air, and then poured and cooled without chemical change.

apply to the tendency to be oxidized by oxygen, chlorine, etc. By thinking again about what you did in Question 2.3, you should be able to imagine how useful such a grading would be in predicting the consequences of putting one metal in contact with the compound or aqueous ion of another. We shall return to this idea in Section 4, but first we shall see how some support for it can be found in the general chemistry of the extraction of metals from their ores.

2.5 Summary of Section 2

1   Oxidation is commonly defined as the loss of electrons, and reduction as the gain of electrons.

2   We have tried to classify metals according to the vigour of their reaction with $H^+(aq)$. In the first class, we included metals like copper, silver and gold, which undergo no reaction; in the second, we included metals like iron, tin and zinc, which react mildly; in the third class, we included metals like the alkali and alkaline earth metals, which react vigorously.

3   It then appears that metals in the third class will reduce the aqueous ions of metals in the first and second classes; likewise, metals in the second class reduce the aqueous ions of metals in the first. However, reactions of the reverse type do not happen.

4   Metals in the third class combine more vigorously with oxygen and chlorine than do metals in the first class.

5   Taken together, these observations suggest that it might be possible to arrange metals in a series in order of their tendency to be oxidized. Such a series would have valuable predictive powers.

QUESTION 2.4

Use the definitions of oxidation and reduction employed in this Book to determine what element, if any, is oxidized and what element, if any, is reduced in the following reactions:

(i) $Sn(s) + Cu^{2+}(aq) = Sn^{2+}(aq) + Cu(s)$

(ii) $Pb(s) + 2Ag^+(aq) = Pb^{2+}(aq) + 2Ag(s)$

(iii) $Cl_2(g) + 2Br^-(aq) = Br_2(aq) + 2Cl^-(aq)$

(iv) $Ba(s) + Cl_2(g) = BaCl_2(s)$

(v) $Ag^+(aq) + Cl^-(aq) = AgCl(s)$

(vi) $Zn(s) + 2AgI(s) = ZnI_2(s) + 2Ag(s)$

(vii) $AlCl_3(s) + 3K(s) = 3KCl(s) + Al(s)$

(viii) $Fe_2O_3(s) + 2Al(s) = 2Fe(s) + Al_2O_3(s)$

(ix) $2FeO(s) + \frac{1}{2}O_2(g) = Fe_2O_3(s)$

(x) $2Fe^{2+}(aq) + Cl_2(g) = 2Fe^{3+}(aq) + 2Cl^-(aq)$

METALS AND THEIR ORES

3

As the behaviour of caesium and gold reveals, a particular metal often responds in a similar way towards quite different oxidizing agents. This observation would be more striking if we were not so familiar with its practical consequences.

We can begin with the jewellery shown in Figure 1.2 (p. 14). Like so many metallic objects found on archaeological sites, it is made of gold. Other metals favoured by ancient craftsmen were silver and copper. Gold and silver metals were often found free, but most of the copper was extracted from ores. Now, many of the copper objects discovered at archaeological sites are heavily corroded (Figure 3.1), but those of silver and, especially, gold have suffered much less.

Figure 3.1
This Byzantine copper coin from Laodiciae in Turkey reveals obvious signs of copper corrosion.

● What connection exists between the last sentence and the one preceding it?

● The fact that copper has to be extracted from its ores suggests that it forms compounds more readily than silver or gold. At the same time, this greater willingness of copper to form compounds accounts for the corrosion of the artefacts.

In other words, the more readily a metal is oxidized, the more difficult it will be to obtain it from its ores. One further example of our ability to grade metals according to their tendency to be oxidized is a qualitative classification in terms of the ease with which we can extract them from their ores. We can distinguish three classes whose composition is very similar to those given in Section 2.2.

1 The so-called 'noble' metals, which seem reluctant to react with other substances. They are often found free or can be obtained by heating their ores in air. This first class includes gold, silver and mercury.

2 More easily oxidized metals, which are often obtained by heating their ores in air to form oxides, then heating the oxides with some form of carbon at furnace temperatures up to about 1 500 °C, and then cooling; for example

$$PbO(s) + C(s) = Pb(s) + CO(g) \qquad (3.1)$$

This second class includes copper, lead, tin, iron and zinc.

3 The third class of metals has a greater tendency to be oxidized than the first two. It includes the alkali metals and the alkaline earth metals, which, because they cannot readily be obtained by heating their compounds with carbon, were unknown as late as 1800. It was the invention of electrical cells and electrolysis that first enabled chemists to extract these metals from their ores.

To reinforce these points, we now consider the extraction and uses of one metal from each of the three classes. The metals we have chosen are mercury, tin and aluminium.

3.1 Mercury

Metallic mercury (Figure 3.2) is a heavy silvery liquid with a density nearly 14 times that of water. In nature, mercury is only rarely found in quantity as the free metal; by far the most common source is cinnabar, mercury sulfide, HgS (Figure 3.3). The metal can be obtained by simply heating the ore in air, in a furnace:

$$HgS(s) + O_2(g) = Hg(g) + SO_2(g) \qquad (3.2)$$

The mercury vapour can then be easily condensed to the liquid metal below its boiling temperature of 357 °C.

(a)

Figure 3.2
Mercury is such a dense liquid that
(a) cannonballs float in it. (b) Floating
human beings seem to barely penetrate
the surface.

(b)

Figure 3.3
Mercury sulfide, cinnabar, being heated in
order to obtain mercury.

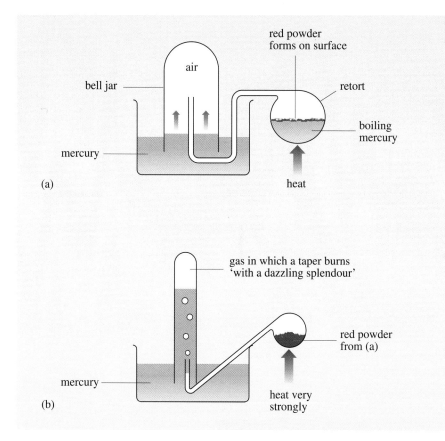

Figure 3.4
The famous experiment of Antoine Lavoisier: (a) mercury is boiled in contact with a sealed volume of air; (b) the red powder thus formed is heated more strongly.

Science has benefited very greatly from the unusual properties of mercury. Chemists in particular recall the famous experiment (Figure 3.4a) of Antoine Lavoisier (1734–1794), which established the nature of combustion and made their subject a modern science. Boiling mercury in the retort was kept in contact with a sealed volume of air in the bell jar and the body of the retort for 12 days. The volume of air decreased by 130 cm³ to about five-sixths of its former value, and a red powder appeared on the surface of the mercury in the retort. Animals placed in the residual air 'suffocated in a few seconds', and a burning taper 'was extinguished as if it had been immersed in water'.

⬤ Use modern chemistry to explain these facts and to identify the red powder.

⬤ The boiling mercury has removed nearly all of the 20% of oxygen in the air by reacting with it and forming a mercury oxide — the red powder.

After the retort had cooled, Lavoisier scraped all of the red powder off the mercury surface and heated it very strongly (Figure 3.4b). A gas was produced and the red powder disappeared, leaving mercury in its place. The volume of gas was 130 cm³. A taper 'burned in it with a dazzling splendour, and charcoal, instead of consuming quietly as it does in common air, threw out such a brilliant light that the eyes could scarcely endure it'. If the sample of new gas was added to the lethal gaseous residue from the experiment of Figure 3.4a, the combination possessed all the properties of ordinary air.

⬤ So what happens in the experiment of Figure 3.4b?

The oxide of mercury decomposes, regenerating the oxygen that had been removed from the air in Figure 3.4a in a pure state. The other product of the decomposition is metallic mercury.

For Lavoisier, these experiments revealed the nature of combustion: a reaction of the burning substance with a component of the air, which he called oxygen. But they also have a broader significance. The first experiment involves *synthesis*: a chemical *compound* is made by the combination of mercury and oxygen. The second involves *analysis*: the compound is split up into mercury and oxygen, two substances that were called chemical elements because they cannot be taken apart in the same sort of way. Lavoisier therefore called the red compound mercury oxide, thus building the distinction between elements and compounds, and the fundamental idea of a chemical element, into the nomenclature of chemistry. Other chemical substances were then brought into this system of nomenclature. As a result, although for the modern chemist, Lavoisier's *The Elements of Chemistry* (1789) may read like an old-fashioned textbook, previous works on the same subject seem unintelligible.

The experiments of Figure 3.4 depend on the delicate equilibrium between mercury and oxygen on the one hand, and mercury oxide (HgO) on the other:

$$2Hg + O_2(g) \rightleftharpoons 2HgO(s) \tag{3.3}$$

At the boiling temperature of mercury (357 °C), equilibrium lies to the right and mercury reacts with oxygen to form HgO. But at 600 °C, equilibrium lies to the left, and there is decomposition into mercury and oxygen (Figure 3.5).

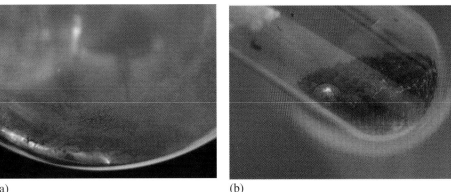

(a)   (b)

Figure 3.5
(a) Red mercury oxide, HgO, being formed on mercury boiling in air; (b) the same oxide undergoing thermal decomposition at a higher temperature to mercury and oxygen

Is this behaviour of mercury oxide consistent with our classification of mercury as a class 1 metal?

Yes; the oxide of mercury yields the metal if it is simply heated.

Because mercury is both a liquid and a good electronic conductor at room temperature, it also has many uses in wiring and switching devices. Another major use of mercury is in the manufacture of chlorine and sodium hydroxide. It is, however, very toxic, and has been implicated in some highly publicized cases of environmental pollution. In 1953, methane-producing microbes in the sludge at the bottom of Minamata Bay in Japan, converted waste mercury metal from a factory into the monomethylmercury ion, $Hg(CH_3)^+(aq)$. In this form, mercury readily penetrates the walls of living cells, and it entered the food chain, ultimately causing brain damage and death among some of those who ate fish caught in the bay. Because of cases like this, the use of mercury has declined in recent years, notably in electrical batteries, and in pesticides and fungicides, where some products have been banned. World production in 1996 was 2 800 tonnes, less than half that in 1986 and only a quarter of the 1971 figure. More than one-third of 1996 production came from the most famous of all cinnabar mines at Almaden in Spain. Figure 3.2 was photographed at this site, which has been in use since Roman times.

3.2 Tin

The chief ore of tin is cassiterite, SnO_2, and this is the only one from which the metal is extracted (Figure 3.6). Currently, the main suppliers of tin ore are China, Indonesia, Peru and Brazil. Much tin is obtained by heating cassiterite with coal in a furnace at 1 200–1 300 °C, when liquid tin and oxides of carbon are formed; for example

$$SnO_2(s) + 2C(s) = Sn(l) + 2CO(g) \qquad (3.4)$$

(a)   (b)

Figure 3.6
Cassiterite, the principal ore of tin (left), and metallic tin (right).

Since the two reactants are solids, intimate contact is inhibited, and the process would be slow if Equation 3.4 were the sole reaction. However, gas/solid reactions are much faster, and the carbon reacts with carbon dioxide, which is always present in the furnace atmosphere:

$$2C(s) + 2CO_2(g) = 4CO(g) \qquad (3.5)$$

A second gas/solid reaction between carbon monoxide and SnO_2 then yields liquid tin:

$$2CO(g) + SnO_2(s) = 2CO_2(g) + Sn(l) \qquad (3.6)$$

⬤ What is the result of adding Equations 3.5 and 3.6?

⬤ It is Equation 3.4, the overall reaction.

Various furnaces can be used, and Figure 3.7 shows the most common type, known as a *reverberatory furnace*. The burner is fired by oil, gas or pulverized coal, and the flame, with its intensely hot combustion products, is reflected or 'reverberates' from the curved roof on to the charge, which is fed in at intervals through pipes in the roof. The charge consists of tin ore, coal and a 'flux' such as lime, which promotes

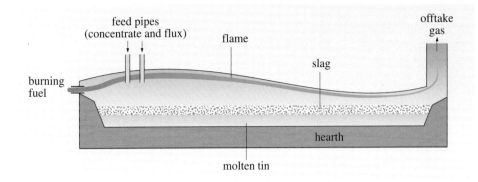

Figure 3.7
A reverberatory furnace of the type used to make metallic tin. A typical furnace is about 30 m long and about 10 m wide.

fusion. The latter forms calcium oxide, CaO, in the furnace; in turn this combines with impurities, such as SiO_2, to yield a liquid slag. The slag floats on the liquid tin, which collects on a hearth at the base of the furnace, where it can be tapped off and solidified. At this stage, the tin contains impurities, such as iron and arsenic. These are much reduced by heating to just above the melting temperature of tin, and drawing the liquid from a solid 'dross', containing impurities with higher melting temperatures.

World production of metallic tin in 1997 was 231 000 tonnes; 70 000 tonnes of it came from China, 40 000 tonnes from Indonesia, and 40 000 tonnes from Malaysia. The South Crofty, the UK's last operating tin mine closed early in 1998, signalling the end of the great Cornish tin industry of the past. Over one-third of tin production is consumed as tin plate in the canning industry, and about a quarter in solders. The melting temperatures of tin and lead are 232 °C and 327 °C, respectively, but a typical general engineering solder for electronics contains 60% tin and 40% lead, and melts well below either temperature in the range 183–188 °C. Standard plumbers' solders, which contain a higher proportion of lead (about 70%) have a higher melting range (183–255 °C).

(a)

3.3 Aluminium

The oxide of aluminium, Al_2O_3, occurs in many parts of the world in combination with between one and three molecules of water. These deposits are called *bauxites* (Figure 3.8a). In the extraction of aluminium, bauxite is first heated under pressure with 4 mol litre^{-1} NaOH solution at 140 °C. This procedure takes advantage of an unusual property of Al_2O_3. It is an oxide that is willing to dissolve in strong bases *as well as* in acids; such an oxide is said to be **amphoteric**. The ion $Al(OH)_4^-$(aq), in which the coordination is tetrahedral (Figure 3.9) is the main product:

$$Al_2O_3(s) + 3H_2O(l) + 2OH^-(aq) = 2Al(OH)_4^-(aq) \qquad (3.7)$$

Insoluble impurities, such as the oxides of iron, silicon and titanium, can then be filtered off. When the solution is cooled and diluted, the equilibrium swings back to the left and Al_2O_3 can be reprecipitated in a hydrated form as the hydroxide, $Al(OH)_3$. If this is heated at 1 100 °C, pure Al_2O_3 is formed:

$$2Al(OH)_3(s) = Al_2O_3(s) + 3H_2O(g) \qquad (3.8)$$

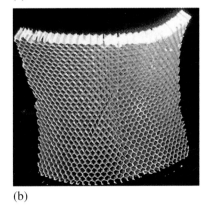

(b)

Figure 3.8
(a) Bauxite, the principal ore of aluminium, is usually red–brown due to the presence of iron minerals; (b) aluminium metal.

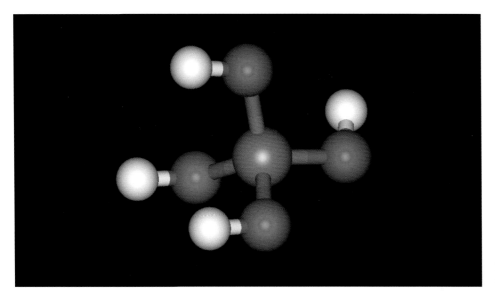

Figure 3.9
The $Al(OH)_4^-$ ion.

The next step is the extraction of the metal from the oxide. Aluminium cannot be obtained by carbon reduction because this yields a carbide of aluminium. Sodium can be obtained by the electrolysis of molten sodium chloride: sodium ions in the melt move to the negative electrode and are deposited as sodium metal. Similarly, if Al_2O_3 is converted into aluminium chloride, $AlCl_3$, one might then hope to obtain aluminium in like fashion. Attempted electrolysis, however, is crowned with disappointment. The fused chloride does not conduct electricity, so the melt evidently contains covalent molecules rather than ions. Owing to these difficulties, the first authenticated samples of metallic aluminium were obtained by a non-electrolytic method, employing reduction of the trichloride with metallic potassium:

$$3K + AlCl_3 = Al + 3KCl \tag{3.9}$$

This type of reaction, using alkali metals, was first operated on an industrial scale by a Frenchman, H. E. St-C. Deville. In 1854, at international exhibitions in Paris, guests of the extrovert Emperor Napoleon III enjoyed the privilege of dining with aluminium cutlery.

At this time, aluminium cost about £50 per pound, because the two reactants in the manufacturing process had to be made from raw materials by rather expensive methods. Moreover, potassium and aluminium trichloride are difficult to handle because they both react with water. Clearly, a new process working directly with the oxide was needed, and the solution was provided almost simultaneously by a Frenchman, Paul Héroult, and an American, Charles Hall, in 1886. Both men were 22 years old at the time.

Hall (Figure 3.10), became interested in the aluminium problem while he was a student at Oberlin College, Ohio. Immediately after graduation, he borrowed a battery, a crucible and a few other bits of apparatus from his professor, and worked on the extraction problem in his father's woodshed. As the oxide fuses at such high temperatures, electrolysis of the pure melt was obviously out, but it occurred to Hall that if he could dissolve the oxide in a suitable solvent, it might then be decomposed electrolytically, just as $CuCl_2$ can be electrolytically decomposed in aqueous solution. After some difficulties, he found his solvent: it was cryolite, Na_3AlF_6, a compound that occurs naturally in large deposits in south-west Greenland.

Figure 3.10
Charles Hall (1863–1914). The son of a protestant clergyman, he studied chemistry at Oberlin College, Ohio where his professor spoke of the fortune that awaited the inventor of an economical method of aluminium extraction. Helped by his elder sister Julia, he converted his father's woodshed into a laboratory, where he successfully operated a miniature version of the modern extraction process in February 1886. Riches soon followed, but not soon enough for his fiancée who became impatient and broke their engagement. Frugal, religious and shy, Hall then devoted himself to the piano and Oberlin College, which received 5 million dollars in his will.

When cryolite is melted at about 1 000 °C and aluminium oxide is dissolved in it, electrolysis with carbon electrodes yields molten aluminium of 99.5% purity at the negative electrode:

negative electrode: $2Al^{3+}(melt) + 6e^- = 2Al(l)$ (3.10)

positive electrode: $3O^{2-}(melt) = \frac{3}{2}O_2(g) + 6e^-$ (3.11)

At the temperature of the furnace, the positive carbon electrodes are burnt away (as carbon monoxide or carbon dioxide gas) by the oxygen formed on their surfaces. Thus, they must constantly be renewed. The cell is kept hot by the electric current through the melt. A typical example is shown in Figures 3.11 and 3.12.

This is the process for aluminium extraction which is still used today. In spite of his success, Hall found difficulty in getting financial support, but in 1889 he arrived in Pittsburgh and persuaded some businessmen to found the Pittsburgh Reduction Company, which subsequently became the gigantic Aluminium Company of America (Alcoa).

Very large amounts of power are required to extract aluminium electrolytically, and, if a choice has to be made, plants are usually located near sources of cheap hydroelectric power rather than close to bauxite deposits. Jamaica has huge sources of bauxite, but exports it to countries such as Canada, where there are two very large plants at Arvida, Quebec, on the Saguenay River, and Kitimat, British Columbia.

For many years, British aluminium production was carried out at two small plants at Kinlochleven and Fort William. In 1970, their combined production was only about 40 000 tonnes. However, in 1968, the British Government provided loans and investment grants to encourage the construction of three large smelters, each with a production capacity of about 100 000 tonnes per annum, at Holyhead, Lynemouth and Invergordon. By 1980 British aluminium production had, in consequence, reached a peak of 374 000 tonnes, but during the economic recession of the early 1980s, the Invergordon smelter was closed, and two years later, production was down to 240 000 tonnes.

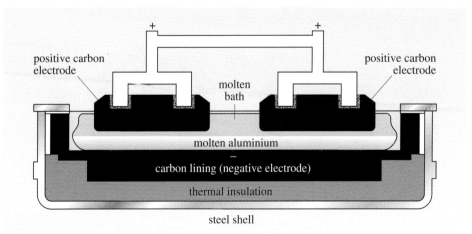

Figure 3.11 Hall electrolytic cell used for aluminium production. A typical plant may contain as many as 1 000 cells, each about 10 m long and 3 m wide.

Figure 3.12 An interior view of an aluminium smelter at Portland, Victoria in Australia.

These events were part of a global pattern. During the early 1980s, the aluminium industry was in crisis, and production underwent a significant redistribution. Across the developed world, in the EU, the USA and especially in Japan, aluminium smelters were closed or mothballed, and production was shifted to countries that were closer to bauxite deposits and had lower energy and/or labour costs. Brazil and Australia increased production some fourfold, Canada by 50%, and Norway became the largest European producer.

Why were Canada and Norway so favoured?

As noted above, cheap hydroelectric power can mitigate the industry's energy demands; these countries have it.

Such changes, and the growth of the Chinese economy, are apparent in Figure 3.13, which shows world aluminium production for 1978–1998 (Figure 3.13a), together with data for some individual regions and countries (Figure 3.13b). In 1998, the world figure was 22.1 million tonnes, British production being 260 000 tonnes.

Currently, canning and, especially, transport (Figure 3.14) are potential growth areas for the aluminium industry. With a density of a little more than a third that of stainless steel, aluminium has been associated with the aerospace industry from the very beginnings of powered flight, the engine casing of the Wright brothers' aircraft being made of it. Even today, aluminium remains the dominant metal in all but the most high-performance aircraft.

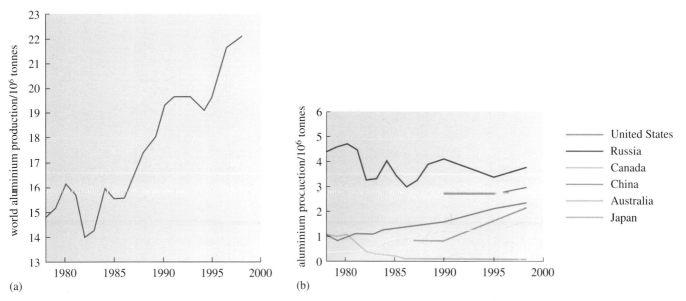

(a)   (b)

Figure 3.13   (a) Recent world aluminium production figures; (b) production data for some individual countries.

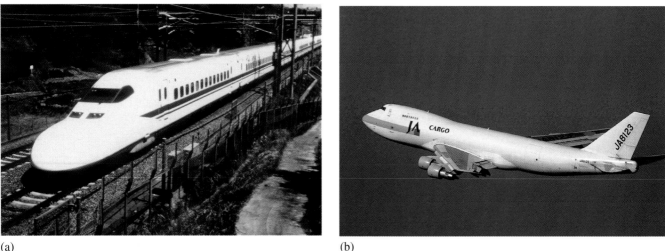

(a)   (b)

Figure 3.14   (a) The Japanese bullet trains with a top speed of $285\,\mathrm{km\,h^{-1}}$ have a double skin of aluminium. (b) Aluminium is the principal structural metal in Jumbo jets; the Boeing 747 contains 75 tonnes.

3.4 Summary of Section 3

1   Our division of metals into three classes according to their tendency to be oxidized correlates well with the relative difficulty of extracting metals from their ores.

2   For example, mercury (in class 1) occasionally occurs naturally, and can also be obtained merely by heating cinnabar, HgS, in air.

3   Tin (in class 2) is obtained by roasting cassiterite, SnO_2, with carbon.

4   Aluminium (in class 3) requires much energy for its extraction, which involves the electrolysis of the pure oxide in molten cryolite (Na_3AlF_6) using graphite electrodes.

METALS AND THEIR EASE OF OXIDATION: A HYPOTHESIS

4

In Sections 2.2–2.4 we examined a series of experiments and observations which suggest that it might be possible to produce a single grading of metals arranged in order of their tendency to be oxidized by various oxidizing agents. This idea is not a new one; it was first propounded in the nineteenth century, and several chemists advanced quite detailed gradings, which were largely based on a more thorough series of qualitative observations than ours. These gradings are often called the **activity series** of metals. A typical example from a textbook of that time is given below:

Cs > Rb > K > Na > Li > Ba > Sr > Ca > Mg
> Al > Zn > Fe > Sn > Pb > H > Cu > Hg > Ag > Au

Hydrogen has also been included in the series because many metals reduce hydrogen ions to hydrogen gas.

○ Is this series consistent with the classifications advanced in Sections 2.2 and 3? If so, where do the dividing lines between the classes come in the series?

○ It is consistent. The dividing lines come between aluminium and zinc, and between hydrogen and copper.

If this single series correctly expresses the relative tendency of metals to be oxidized by any oxidizing agent, it could be used to predict the outcome of many metal reactions. Whereas before we assigned metals to one of three classes on the basis of their tendency to be oxidized, we now have a series that grades the metals within the classes.

Test whether you have properly appreciated this by doing Question 4.1 now.

QUESTION 4.1

By assuming that the classification or grading of metals made in Sections 2–4 correctly represents 'the tendency of metals to be oxidized' by a variety of oxidizing agents, predict the outcome, if any, of an attempted reaction in which the following pairs of substances are heated together. Write balanced equations if you think reactions will occur. (In most cases we have not included physical states because they will vary according to the temperature of the reaction.)

(i)   Cu and ZnI_2

(ii) Zn and CuO $= ZnO + Cu$

(iii) Zn and $FeCl_2$ $= ZnCl_2 + Fe$

(iv) Fe and KI

(v)   3 K and $AlCl_3$ $= KCl_3 + Al$

(vi) Hg and Al_2O_3

(vii) Zn(s) and 2 Ag^+(aq) $= Zn^{2+}_{(aq)} + 2Ag(s)$

(viii) Fe and $PbBr_2$ $= FeBr_2 + Pb$

(ix) Zn(s) and H_2O(g) $= ZnO_{(s)} + H_2(g)$

(x)   Ba and $FeBr_2$ $= BaBr_2 + Fe$

You should now do Activity 4.1.

ACTIVITY 4.1 Testing an activity series

In this activity, the fourth sequence of *Reactions of metals*, on the CD-ROM associated with this Book, you will see metals added to solutions of the salts of other metals. You are asked to decide whether a reaction occurs. The reactions are performed first in the cold and then, in some cases, warmed to speed up any reaction. As most of the solutions are colourless, you will need to make especially careful observations of any changes in the nature of the metal. Blackening or swelling are frequent signs of reaction when a new metal is precipitated or formed on the surface of another.

The following reactions are attempted:

(i)   iron filings and copper sulfate solution;

(ii)   copper and silver nitrate solution;

(iii)   magnesium and zinc sulfate solution;

(iv)   magnesium and lead nitrate solution;

(v)   aluminium and copper sulfate solution;   see G.13.2

(vi)   aluminium and dilute hydrochloric acid;

(vii)   zinc and lead nitrate solution.

You will find that some metals cause evolution of hydrogen gas from the solution, especially on warming. When you have finished, consider whether your observations are consistent with the implications of the series given at the beginning of this Section. Explain why you think they are consistent or not consistent as the case may be. For a comment on the experimental results, see p. 40.

Before beginning the next Section, attempt Question 4.2.

QUESTION 4.2

The dissociation of the weak acid, acetic acid, in aqueous solution, may be written

$$HAc(aq) = H^+(aq) + Ac^-(aq) \qquad K = \frac{[H^+_{aq}][Ac^-_{aq}]}{[HAc_{aq}]} = 1.8 \times 10^{-5} \text{ mol litre}^{-1} \qquad (4.1)$$

Write down an expression for the equilibrium constant of the reaction.

4.1 A critical look at the hypothesis

It is now time to examine critically our proposal that a single grading of metals can correctly express their tendency to be oxidized by various oxidizing agents. First then, let us turn to the language that we have used. Words and phrases such as 'tendency to be oxidized', 'willingness to be oxidized', 'violence of reaction' and 'reactivity' do not describe quantitative concepts that can be precisely defined and expressed in measurable units. This is a serious limitation, which hinders extension of the ideas if they are correct, and prevents the uncovering of their weaknesses if they are not. Until now, our observations have not been quantitative.

If you think carefully about some of the observations that we have made, you may already be able to spot weaknesses in the arguments. Consider, for example, the classification in Section 2.2, which was built on observations made in that Section and supported by others in Section 2.3. In Section 2.3 we used the fact that the equilibrium position in the reaction

$$Zn(s) + Cu^{2+}(aq) = Zn^{2+}(aq) + Cu(s) \qquad (2.12)$$

lay to the right-hand side as an indication that zinc was more readily oxidized than copper.

⬤ What criterion did we use in Section 2.2 to show that magnesium was more readily oxidized than zinc or tin by aqueous hydrogen ions?

⬤ We used the vigour, or to be more precise, the *rate* of the reaction.

Now you know the importance of distinguishing the concept of how far a reaction goes from how fast it goes, that is, of distinguishing the equilibrium constant from the rate of reaction. The two are often completely unconnected; for example, a reaction like the combustion of natural gas, which has a huge equilibrium constant, proceeds immeasurably slowly at room temperature (in the absence of a spark!). Yet now we are trying to grade metals according to their tendency to be oxidized, by considering at the same time both rates and positions of equilibrium. The idea of grading the metals according to qualitative observations makes an immediate appeal to the senses, but it may be that we have allowed the visual force of the idea to bring us into conflict with important chemical principles.

What we badly need, is an approach in which the reactions of metals are compared quantitatively under the same conditions in terms of either rates or positions of equilibrium. Which of these two approaches is likely to be the more helpful? We can answer this by noting that, in developing our ideas, we have discussed several reactions that do not occur: reactions such as

$$Cu(s) + 2H^+(aq) = Cu^{2+}(aq) + H_2(g) \qquad (2.9)$$
$$Cu(s) + Mg^{2+}(aq) = Cu^{2+}(aq) + Mg(s) \qquad (4.2)$$

Clearly, we cannot determine the rates of these reactions if we cannot detect their occurrence. This suggests that an approach in which we compare the reactions of metals by their rates would be quite unproductive in certain important cases.

This leaves us with the approach in which we compare equilibrium constants. The expression that we wrote for an equilibrium constant, K, in Question 4.2 suggests that values of K can be determined by finding the concentrations of the reactants and products in the equilibrium state. However, Reaction 4.2 occurs to such a small extent that it is not possible to determine the concentration of $Cu^{2+}(aq)$ — written $[Cu^{2+}(aq)]$ — by orthodox chemical analysis. You may therefore raise the same objection to the comparison of equilibrium constants that we raised to the comparison of rates: how can we compare the equilibrium constants of reactions that do not appear to occur? As we shall show you between this point and the end of Section 11, the science of chemical thermodynamics enables us to do just that. For this reason, we now embark on an approach in which we compare, under the same conditions, the equilibrium positions of the reactions of metals.

4.2 Summary of Section 4

1   The methods and observations that we have used to arrange metals in order
    of their tendency to be oxidized are not quantitative; nor do they make the
    necessary distinction between the equilibrium constant and the rate of a
    chemical reaction.

2   A comparison of metal reactivity based on equilibrium constants is likely to
    be more productive than one based on reaction rates.

STUDY NOTE

All reactions in Activity 4.1 have the expected outcome except reaction (v). This
exception is discussed in Question 13.2.

EQUILIBRIUM: A RESTATEMENT OF THE PROBLEM

5

Let us start by trying to formulate the problem in a more precise way. One way of recognizing equilibrium in a chemical system is by the 'appearance of quiescence': nothing changes as time passes. Once at equilibrium, or in an equilibrium state, a chemical system does not change with time unless disturbed in some way: thereafter it heads inexorably back toward equilibrium.

With this in mind, consider ordinary table salt, sodium chloride. This is a compound of sodium and chlorine. However, under normal conditions, the table salt you use at home does not decompose into sodium metal and chlorine gas of its own accord. (Perish the thought!)

To be more precise, at 298.15 K (25 °C) the equilibrium position for the reaction

$$NaCl(s) = Na(s) + \tfrac{1}{2}Cl_2(g) \qquad\qquad (5.1)$$

evidently lies so far over to the left-hand side that it effectively does not happen at all. In other words, solid sodium chloride (rather than a mixture of sodium and chlorine) corresponds to the equilibrium state for this chemical system.

Further evidence is provided by the experimental observation that the reverse of Equation 5.1, the reaction between metallic sodium and chlorine,

$$Na(s) + \tfrac{1}{2}Cl_2(g) = NaCl(s) \qquad\qquad (5.2)$$

happens spontaneously: the equilibrium state again corresponds to NaCl.

So far so good. Seen in a broader context, however, the chemical system described above shares a feature in common with other naturally occurring (or spontaneous) processes: it has a direction. For example, the spontaneous reaction in Equation 5.2 behaves rather like a stone rolling downhill: there is a 'downhill' character to this reaction that corresponds to the movement toward equilibrium.

Figure 5.1
Inanimate objects do not roll uphill!

By contrast, there is an 'uphill' character to its reverse — Equation 5.1. Just as balls, etc., don't roll uphill on their own (Figure 5.1), so too the movement away from equilibrium in this chemical system can be achieved only with the help of outside assistance, for example, by melting the sodium chloride and electrolysing it.

These ideas apply equally well to all chemical reactions. Any spontaneous reaction moves toward, never away from, equilibrium. But what is the underlying *reason* for this behaviour? What determines the equilibrium state in a chemical system, and hence governs whether a reaction can happen or not?

THOMSEN'S HYPOTHESIS: TOWARDS A SOLUTION?

(a)

One way towards an answer to the question posed at the end of Section 5 is suggested by the analogy we used earlier: when a stone rolls downhill, we explain its motion by saying that it happens because the stone thereby achieves a state of lower gravitational (or potential) energy. Is there a model here for chemical reactions? Perhaps a chemical reaction can occur only if the products are of lower energy than the reactants?

Reactions are classified as either *exothermic* (releasing energy to the surroundings as heat) or *endothermic* (absorbing energy from the surroundings as heat). The immediate effect of an exothermic reaction is generally a rise in the temperature of the reaction mixture. For instance, when aluminium is dropped into a dish of liquid bromine, a reaction occurs and the dish gets hot (Figure 6.1).

But common laboratory glassware, like a beaker or test-tube, is a rather poor heat insulator: as soon as the temperature rises above that of the surroundings, heat begins to escape from the reaction vessel. This simple observation points to the one universal statement that can be made about exothermic reactions: if the products of such a reaction are *finally* obtained *at the same temperature* as the reactants, then heat must have left the reaction vessel. This, and the complementary case of an endothermic reaction, both at constant temperature[*], are illustrated schematically in Figure 6.2.

(b)

Figure 6.1
(a) Aluminium foil immediately after addition to liquid bromine.
(b) A violent exothermic reaction has begun: much heat is evolved, along with fumes and light.

With the additional proviso that the reaction also takes place at constant pressure[†], the heat (q) released or absorbed by a reaction at constant temperature can then be identified with the **enthalpy** (from the Greek *en*, meaning 'in', and *thalpē*, meaning 'make hot') **of reaction**, ΔH, at that temperature.

Thus,

$$\Delta H = q \text{ (at constant } T \text{ and } p)\qquad(6.1)$$

Remember that the symbol Δ (Greek capital 'delta') means a *change* in some physical quantity, in this case a change in enthalpy, H. So, for a reaction at constant T and p,

$$\Delta H = H(\text{products}) - H(\text{reactants})\qquad(6.2)$$

To summarize: Equations 6.1 and 6.2 together imply that the heat released (or absorbed) by a reaction is a measure of the difference in energy between reactants and products *under the same conditions of temperature and pressure*.

---

[*] Throughout this Book, the phrase 'at constant temperature' simply implies that the initial and final temperatures are the same (as shown in Figure 6.2), no matter what happens in between. It does not necessarily imply that the temperature is actually held constant *throughout* the reaction, as in a thermostat for example, although this may indeed be so in certain cases.

[†] This condition derives from the fact that laboratory reactions are usually carried out in vessels open to the atmosphere, that is, at constant (atmospheric) pressure. For this reason, chemists chose to define a property, *enthalpy*, which can be used to express the energy transferred under this condition. A further comment on conditions governing the validity of Equation 6.1 is discussed in the Appendix.

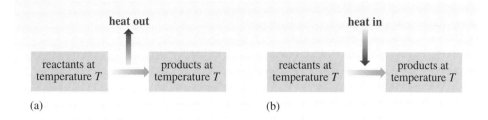

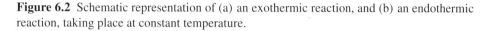

(a)   (b)

Figure 6.2 Schematic representation of (a) an exothermic reaction, and (b) an endothermic reaction, taking place at constant temperature.

○ What are the signs (i.e. positive or negative) of ΔH for an exothermic reaction and for an endothermic reaction?

○ This derives from the sign convention for heat transfer: q is positive (> 0) if heat is absorbed by the system of interest (the reaction mixture in this case), but negative (< 0) if heat is released to the surroundings. *From Equation 6.1, therefore, ΔH is negative for an exothermic reaction and positive for an endothermic one.*

In the context of Equation 6.2 therefore, only exothermic reactions result in products of *lower* energy than the reactants. (This is shown schematically in Figure 6.3a.) The energy analogy with the motion of a stone, proposed at the start of this Section, therefore implies that only exothermic reactions can happen spontaneously, whereas endothermic reactions (for which the products are of higher energy than the reactants, Figure 6.3b) *cannot* occur on their own. In 1853, this hypothesis was proposed by the Danish chemist, Julius Thomsen (Figure 6.4).

Figure 6.4
Julius Thomsen (1826–1909). This photograph captures the solitary nature of this Danish chemist, who determined many important enthalpies of reaction, while refusing frequent offers of collaboration from others. This partly explains the accuracy of his values, because foreign hands were never able to tamper with the internal consistency of his work. He also published (1895) an early long form of the Periodic Table (subsequently improved by Alfred Werner), which was used by Niels Bohr, his fellow countryman, to deduce the electronic configurations of atoms. For this, Bohr won the Nobel Prize for Physics in 1922.

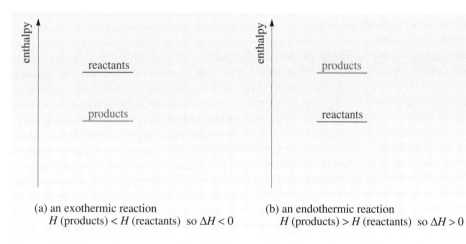

(a) an exothermic reaction
H (products) $<$ H (reactants) so $\Delta H < 0$

(b) an endothermic reaction
H (products) $>$ H (reactants) so $\Delta H > 0$

Figure 6.3 Schematic representation of (a) an exothermic reaction, and (b) an endothermic reaction, defined in terms of enthalpy changes at constant temperature and pressure.

You should now try the following Question, to see how this hypothesis compares with the observed behaviour of a chemical system.

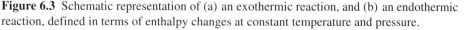

43

QUESTION 6.1

Consider the reaction between sodium and chlorine discussed in Section 5:

$$Na(s) + \frac{1}{2}Cl_2(g) = NaCl(s) \qquad\qquad (5.2)$$

If one mole of sodium reacts with half a mole of chlorine at 298.15 K (25 °C), the enthalpy change is:

$$\Delta H = -411.2 \text{ kJ}$$

(a) Is the reaction exothermic or endothermic? What is the value of ΔH for the reverse reaction, Equation 5.1?   *Exothermic*, $NaCl(s) = Na(s) + \frac{1}{2}Cl_2(g)$; $\Delta H = +411.2\,kJ$

(b) Could Thomsen's hypothesis provide a satisfactory explanation for the observed behaviour of this system, as outlined in Section 5?

Seen in a more critical light, however, just *one* example of a *spontaneous endothermic* process would be sufficient to discredit Thomsen's hypothesis. At room temperature, it is quite difficult to find such examples, but not impossible. One mole of table salt will dissolve spontaneously in a large amount of water:

$$NaCl(s) = Na^+(aq) + Cl^-(aq); \Delta H = +3.9 \text{ kJ at } 298.15 \text{ K} \qquad (6.3)$$

Similarly, the use of thionyl chloride ($SOCl_2$) as a common dehydrating agent in the laboratory depends on another equally spontaneous reaction; when one mole of each reactant is involved, we have:

$$SOCl_2(l) + H_2O(l) = SO_2(g) + 2HCl(g); \Delta H = +47.1 \text{ kJ at } 298.15 \text{ K} \qquad (6.4)$$

Evidently, chemical reactions do not 'roll downhill' on a simple 'enthalpy landscape'. Nevertheless, as the example in Question 6.1 suggests, there is some merit to Thomsen's hypothesis. *Under ambient conditions, the great majority of spontaneous reactions are exothermic.* Because of this, it seems reasonable to suggest that the enthalpy change for a reaction is *one* factor that influences the equilibrium state. But, given the examples cited above, we shall expect to find that it is not the only one.

6.1 Summary of Sections 5 and 6

1   The enthalpy change, ΔH, is negative for an exothermic reaction and positive for an endothermic one. At normal temperatures, most spontaneous reactions are exothermic (negative ΔH).

2   This raises the prospect of some form of energy change that is always negative for a spontaneous reaction. In contrast to the predictions of Thomsen's hypothesis, ΔH will be an important, but not the only, contributor to this energy change.

THE SECOND LAW OF THERMODYNAMICS: THE SOLUTION

7

The discussion in the previous Section brings us back to our original question, but with a little more insight into the sort of answer to expect. As we said in Section 5, the problem is one of predicting the *direction* of spontaneous change in a chemical system. Questions like this lie within the province of one of the most important and far-reaching laws in the whole of science. You may have heard of it: it is called the **second law of thermodynamics**. As you will see, it is this law that provides a criterion for equilibrium in a chemical reaction.

Perhaps the most commonly expressed form of the second law of thermodynamics is the one proposed by the German scientist Rudolf Clausius in 1865: 'the entropy of the universe tends towards a maximum'. Unfortunately, Clausius's famous aphorism is a little too unrestrained to be useful, so we shall adopt the following statement:

> When a natural (or spontaneous) process occurs in an isolated system, the total entropy change within the system is positive.

This means that if, in the world around us, something happens in a region which, during the time-scale of the event, we can regard as an 'isolated system', then,

$$\Delta S_{\text{total}} > 0 \tag{7.1}$$

where Δ again means 'a change', and S denotes the **entropy** (from the Greek *en*, meaning 'in' and *tropē* meaning 'transformation').

We shall treat the second law as one of the 'rules of the game' of thermodynamics — that is, as a basic assumption that can be vindicated only by experiment. Our faith is not misplaced! Deductions from the second law have *never yet* been shown to be invalid, and this success has impressed greater minds than ours (Figure 7.1).

Before going further, however, notice that the above formulation raises two immediate questions: What is an isolated system? What is entropy?

To deal with the first: the word **system** simply means the collection of materials, chemicals, etc., that we are interested in. Once the system is defined, everything else is the surroundings. An **isolated system** is then one that can exchange neither energy (especially heat) nor matter with its surroundings. For example, a reaction mixture enclosed in a sealed, perfectly insulated container (a sort of ideal vacuum flask) would be an isolated system.

🔵 Are these the conditions under which most chemical reactions are carried out?

🔵 No; most laboratory reactions are carried out in poorly insulated containers, often open to the atmosphere. Under these conditions, reactions readily exchange heat (determined by the sign and magnitude of ΔH) with their surroundings.

Figure 7.1
Albert Einstein (1879–1955). A loose translation of one of his many quotable remarks runs as follows. 'The most impressive theories are those which combine very simple premises with a very broad range of applications, and which establish relationships between many things that might otherwise stay unconnected…. I am convinced that thermodynamics is the only universal physical theory that will never be overthrown.'

Thus, reacting chemicals, *by themselves*, cannot normally be regarded as an isolated system. So as you will see, to apply the second law to a chemical reaction carried out under normal laboratory conditions, our definition of the system must extend beyond the boundaries of the chemicals.

But what about the second, more fundamental question: what is entropy?

7.1 Entropy

The question that we have just posed can be approached in two very different ways. Historically, the concept of entropy emerged as a natural development of thermodynamics. Despite its central importance as the arbiter of the direction of spontaneous change, from this point of view, entropy is just another physical quantity — like volume or mass, for example. In other words, entropy can be considered as a property of matter in bulk — just like the more familiar properties mentioned above. The question 'What is entropy?' then resolves itself into the more practical question 'How is entropy measured?' In this sense, the concept of entropy would be meaningful even if we knew nothing about atoms or molecules.

On the other hand, it is possible to start from a theoretical *model* of how atoms and molecules behave, and hence build up a 'molecular' picture of entropy and entropy changes. Unfortunately, the connection between entropy and molecular properties is not at all obvious. However, it is in this context that you may have heard entropy associated with 'chaos' or 'disorder' or 'mixed-upness' or words of a similar sort. This can be misleading, because words like disorder have a rather special meaning in this context — a meaning that is not obviously connected with their everyday sense. One simple example must suffice. At 298.15 K and 1 atm pressure, Avogadro's law tells us that equal volumes of gases, such as the noble gases helium and xenon, contain the same number of molecules (Figure 7.2). The noble gases are monatomic; that is, the molecules are atoms. It is found experimentally (see Section 10.3) that the entropy of the xenon gas sample is roughly 50% greater than that of the helium. But there is no obvious sense in which the atoms in the xenon sample are the more disordered or 'disarranged'.

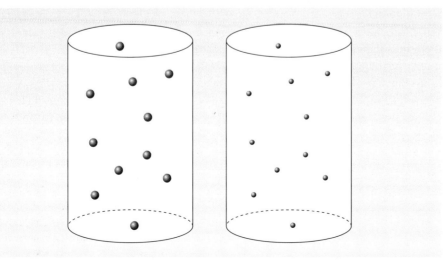

Figure 7.2
Equal volumes of helium and xenon at the same temperature and pressure contain the same number of molecules — in this illustration, eleven! The xenon molecules (left) are larger and more massive than those of helium (right), but there is no immediately obvious sense in which the xenon sample is more disordered.

From now on, we shall restrict ourselves to the 'thermodynamic' approach and ask how entropy is measured. For this purpose, we ask you to accept the following, essentially operational, **definitions of the entropy change** in three very important cases (all at constant temperature and pressure):

1   When heat is transferred to or from a pure substance at a temperature T, the change in the entropy of the substance is

$$\Delta S_{\text{substance}} = \frac{q}{T} \qquad (7.2)$$

where, as usual, the value of q is positive if heat is absorbed by the substance.

2   When heat is transferred to or from a large heat reservoir at a temperature T, such as the atmosphere around us, the change in the entropy of the reservoir is given by

$$\Delta S_{\text{reservoir}} = \frac{q}{T} \qquad (7.3)$$

3   When a chemical reaction takes place, the change in entropy during the reaction is given by

$$\Delta S_{\text{reac.}} = S(\text{products}) - S(\text{reactants}) \qquad (7.4)$$

⚪   According to the definitions in Equations 7.2 and 7.3, what is the SI unit of entropy?

⚪   The SI units of heat and temperature are the joule (J) and the kelvin (K), respectively. So entropy has the unit joule per kelvin ($J\,K^{-1}$).

These three cases have been chosen carefully. Case 1 is, as you will see in Section 10.3, used to determine entropies of pure substances. When these have been obtained, case 3 can be used to determine entropy changes in chemical reactions. Finally, case 2 will enable us to discover a *new* function that does the job we require; that is, it determines the direction of spontaneous change in a chemical system at constant temperature and pressure.

7.2 The direction of heat flow

To gain some familiarity with the second law, and with case 1 (Equation 7.2), let us see if they are consistent with an everyday observation.

Suppose you have two large blocks of iron — block 1 and block 2, say — at temperatures T_1 and T_2, respectively. Suppose further that $T_2 > T_1$, and that you then touch the blocks together briefly so that a small amount of heat passes from one to the other. Finally, assume that the heat transferred is small enough to leave the two temperatures essentially unchanged.

⚪   In which direction does heat flow?

⚪   From block 2 to block 1 — from the block at the higher temperature (T_2) to the block at the lower temperature (T_1).

Suppose that q is the amount of heat transferred in this way (Figure 7.3).

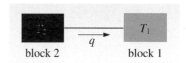

Figure 7.3
Two large metal blocks at temperatures T_1 and T_2 ($T_2 > T_1$) are briefly allowed to touch, or connected by a thin wire, so that a small amount of heat, q, passes from the block at T_2 to the block at T_1.

47

• What are the values of ΔS for block 1 and block 2, respectively?

• Block 1 *absorbs* heat, q, so according to Equation 7.2, $\Delta S_1 = q/T_1$. Block 2 loses an identical amount of heat, so $\Delta S_2 = -q/T_2$.

• If the two blocks are considered as *one* system, what is the total entropy change?

• $\Delta S_{total} = \Delta S_1 + \Delta S_2 = q/T_1 - q/T_2$ (7.5)

• Taking the two blocks as an isolated system (that is, assuming no heat loss to the surroundings), is the heat transfer in accord with the second law?

• Yes; because T_2 is greater than T_1, q/T_2 is less than q/T_1, so ΔS_{total} is positive.

QUESTION 7.1

Repeat this treatment by considering the case in which heat passes from block 1 at the lower temperature T_1 to block 2 at the higher temperature T_2. Show that *this* process *violates* the second law of thermodynamics.

In summary: our formulation of the second law, and our definition of entropy change in case 1, leads to the idea that heat *cannot* pass spontaneously from a colder to a hotter body, which is in accord with everyday experience.

• Can you see why this must be so?

• According to the discussion above, transfer of heat from a colder to a hotter body would be accompanied by a forbidden *decrease* in entropy.

Indeed, this statement is an alternative formulation of the second law.

Examples like those discussed in this Section reveal the difference between entropy and energy.

> Energy is conserved; it can neither be created nor destroyed. But entropy is not like that: it increases, or is created, in all natural (or spontaneous) processes.

Let us now see if the second law can provide the condition that allows a chemical reaction to occur.

7.3 The direction of chemical change

As an example of chemical change, consider again the reaction between sodium and chlorine:

$$Na(s) + \tfrac{1}{2}Cl_2(g) = NaCl(s) \tag{5.2}$$

Let us suppose that this reaction takes place at a constant pressure of 1 atm in a large laboratory. Suppose further that we start with one mole of sodium and half a mole of chlorine at a fixed temperature of 298.15 K (25 °C), and eventually finish up with one mole of sodium chloride at the same temperature.

Now, when the reaction takes place, there will be an associated change in entropy, $\Delta S_{reac.}$, given by case 3 (Equation 7.4). As we hinted above, this can be determined from experimental measurements (discussed in Section 10.3): at 298.15 K and 1 atm pressure, it has the following value for the change that we have just described (Reaction 5.2):

$$\Delta S_{reac.} = -90.7 \, \text{J K}^{-1}$$

At first sight, this value seems to be in violation of the second law: the reaction happens, but the entropy decreases! This apparent failure is a reminder of the warning we sounded at the beginning of this Section. Under normal conditions, the reaction mixture is *not* an isolated system: it affects the surroundings to an extent determined by the sign and magnitude of the associated enthalpy change.

Now, the value of ΔH for this reaction was given in Question 6.1:

$$\Delta H = -411.2 \, \text{kJ}$$

So the reaction is exothermic: it releases heat. It follows that the immediate surroundings (the laboratory) have no option but to *absorb* heat, and hence (according to Equation 7.3) to increase in entropy.

Suppose now that, on the time-scale of the reaction, we define our system as the *laboratory*, and regard it as isolated: no net heat is transferred to it or withdrawn from it by the outside world. According to the discussion above, there are then *two* contributions to the total entropy change in this system when the reaction occurs: the change due to the reaction ($-90.7 \, \text{J K}^{-1}$), *and the change in the entropy of the surrounding laboratory*, $\Delta S_{lab.}$. We can obtain an expression for $\Delta S_{lab.}$ by using an equation from Section 6. At constant T and p,

$$\Delta H_{reac.} = q \tag{6.1}$$

Now, the heat evolved by the chemicals is *absorbed* by the surrounding laboratory, so

$$q_{lab.} = -q = -\Delta H_{reac.} \tag{7.6}$$

Thus,

$$\Delta S_{lab.} = \frac{q_{lab.}}{T} = -\frac{\Delta H_{reac.}}{T} \tag{7.7}$$

Notice that because Reaction 5.2 is exothermic, $\Delta H_{reac.}$ is negative. Thus, according to Equation 7.7, $\Delta S_{lab.}$ must be positive, which agrees with the fact that the laboratory absorbs heat.

⬤ What is the total entropy change for the reacting chemicals plus the surrounding laboratory?

⬤ $$\Delta S_{total} = \Delta S_{reac.} + \Delta S_{lab.} = \Delta S_{reac.} - \frac{\Delta H_{reac.}}{T} \tag{7.8}$$

Now according to the second law of thermodynamics, the reaction can occur only if $\Delta S_{total} > 0$. According to Equation 7.8, at constant T and p, this condition is:

$$\Delta S_{reac.} - \frac{\Delta H_{reac.}}{T} > 0 \tag{7.9}$$

This result is illustrated schematically in Figure 7.4.

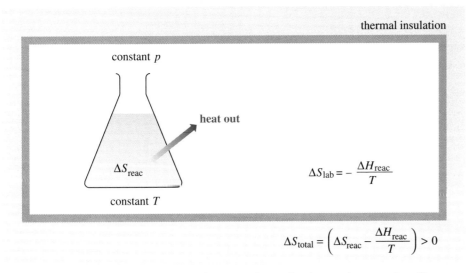

thermal insulation

constant p

heat out

ΔS_{reac}

constant T

$\Delta S_{lab} = - \dfrac{\Delta H_{reac}}{T}$

$$\Delta S_{total} = \left(\Delta S_{reac} - \frac{\Delta H_{reac}}{T} \right) > 0$$

Figure 7.4   The components of the total entropy change for the reaction vessel *and* its surroundings when a spontaneous exothermic reaction occurs at constant temperature and pressure.

QUESTION 7.2

Check that the inequality 7.9 holds for the reaction discussed in this Section:

$$Na(s) + \tfrac{1}{2}Cl_2(g) = NaCl(s) \tag{5.2}$$

Remember that ΔH (= −411.2 kJ) is given in *kilojoules*, that ΔS (= −90.7 J K^{-1}) is given in *joules* per kelvin, and that T = 298.15 K. Is the second law vindicated?

The inequality 7.9 is usually rearranged slightly, by moving the two contributions to the right-hand side and multiplying through by T. If we then drop the subscript 'reac.', this gives $0 > \Delta H - T\Delta S$; or, rearranging,

$$\Delta H - T\Delta S < 0 \tag{7.10}$$

Thus, the quantity $(\Delta H - T\Delta S)$ does for chemical reactions (at constant T and p) what potential energy does for the motion of a ball. As we anticipated in Section 6, ΔH contributes to it, but there is also a second contribution, $-T\Delta S$. For a reaction to occur spontaneously, the quantity $(\Delta H - T\Delta S)$ must be *negative*. Conversely, a positive value for $(\Delta H - T\Delta S)$ corresponds to a movement *away from* equilibrium: such a process never happens of its own accord.

All that remains, then, is to find out how to determine ΔH and ΔS for a given process at a specified temperature: we can then use the inequality above to *predict* if a reaction can happen. We deal first with the more familiar quantity, ΔH.

7.4 Summary of Section 7

1   The direction of spontaneous change is prescribed by the second law of thermodynamics:

> When a natural (or spontaneous) process occurs in an isolated system, the total entropy change within the system is positive; that is, $\Delta S_{total} > 0$.

2   For our purposes, the most important point about an isolated system is that it cannot exchange heat with its surroundings. Thus, reactions at constant temperature and pressure are *not* isolated systems.

3   A change in the entropy of a substance (at constant T and p) can be defined in terms of heat transfer, q, as $\Delta S = q/T$.

4   When a reaction occurs at constant T and p, there are two contributions to ΔS_{total}: the entropy change associated with the reaction itself,

$$\Delta S = S(\text{products}) - S(\text{reactants}) \tag{7.4}$$

and the entropy change of the surroundings, $-\Delta H/T$.

5   According to the second law, the sum of these entropy changes must be positive: this leads to the central inequality for a chemical change at constant T and p, namely: $\Delta H - T\Delta S < 0$.

QUESTION 7.3

At 298.15 K and 1 atm pressure, the values of ΔH and ΔS for the reaction of one mole of thionyl chloride with water

$$SOCl_2(l) + H_2O(l) = SO_2(g) + 2HCl(g) \tag{6.4}$$

are as follows:

$\Delta H = +47.1 \text{ kJ}$

$\Delta S = +334.6 \text{ J K}^{-1}$

Calculate:

(i)   the entropy change of the surroundings;

(ii) the total entropy change, ΔS_{total}. Is the sign of ΔS_{total} in accord with the observation that this reaction happens under the conditions cited above? What is the necessary requirement for a spontaneous endothermic reaction?

THE FIRST LAW OF THERMODYNAMICS

8

In Section 6, we identified ΔH for a reaction with the heat released or absorbed in order to maintain a constant temperature. But so far we have not considered how to measure heat, nor even given a very precise definition of it. To redeem this failing requires another 'rule of the game' — the *first law of thermodynamics*. You will be familiar with this law under the guise of 'the law of conservation of energy'. Our first step will be to formulate this principle in a slightly different way, and then to show that it provides a route to the measurement of enthalpy changes. We shall start with a simple and familiar example.

8.1 The enthalpy change for a pure substance

Suppose that you have a beaker of water (100 g, say) at room temperature and pressure, and you want to raise its temperature — to bring about the change shown schematically in Figure 8.1.

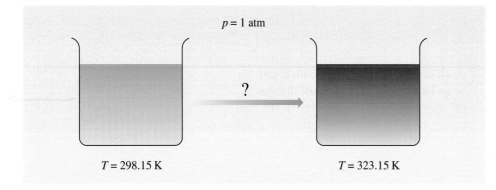

Figure 8.1
Raising the temperature of a sample of water from 298.15 K (25 °C) to 323.15 K (50 °C).

The obvious thing to do is to heat the water, and you could do this in a variety of ways. For example, you could stand the beaker on an electric hotplate, or over a flame, or even in front of an electric fire. Or you could simply immerse it in a bath of hotter water. All these methods depend on there being a *temperature difference* between the system of interest (your beaker of water) and the surroundings (the hotplate, flame, etc.). This, in turn, suggests a more precise definition of **heat**: it is energy transferred because, *and solely because*, of a temperature difference between a system and its surroundings.

● Now think of a different way of bringing about the change in Figure 8.1 — a way that does *not* depend on a temperature difference.

● You may have come up with several possibilities, but the simplest alternative is the one you use every time you switch on an electric kettle or an immersion heater.

At first sight this may not seem too different from our original suggestion — of using a hotplate, for example. After all, an electrical heating element is involved in both cases and the temperature of the water rises because of the higher temperature of the heating element with which the water is in contact. The crucial distinction is illustrated in Figure 8.2: as you can see, *it arises from defining the system so as to include the heating coil.* Energy transfer does *not*, therefore, depend on a temperature difference *between the system and its surroundings.*

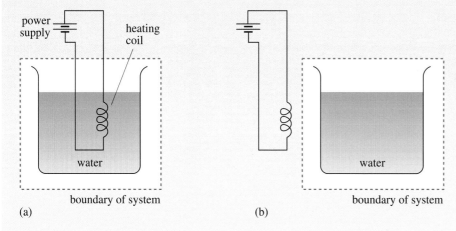

Figure 8.2
Diagram to illustrate the difference between heating a system (b) and doing electrical work on it (a).

The general name for this second sort of energy transfer is **work**, denoted by the symbol w (**electrical work**, w_{el}, in the example cited above[*]). This is the only type of work that we shall consider here.

Thus, it seems that we can transfer energy to the water in the beaker either by heating it directly, or by doing electrical work on it; or we could do both together. Whatever method we choose, there will be an increase in the energy of the system, corresponding to heat and/or electrical work that has caused the change. If the change takes place at constant pressure (as it would in a beaker open to the atmosphere), we again identify this increase in the energy of the system with a change in the enthalpy. This leads to the following general expression:

$$\Delta H = q + w_{el} \text{ (at constant pressure)} \qquad (8.1)$$

where q is the heat *added* to the system (as before), and w_{el} is the electrical work *done on* the system.

Thus, w_{el} is defined as positive if electrical work is done on the system, and energy is thereby transferred to it.

Equation 8.1 is our reformulation of the principle of energy conservation, in terms of the *enthalpy* of the system. As such, it is a special and strictly restricted case of the **first law of thermodynamics** — restricted in that it applies *only* to changes at constant pressure, where the only type of work considered explicitly is electrical

---

[*] At first sight the name 'work' may seem a trifle bizarre in this context: it's not easy to envisage electricity 'doing work'! The term is a legacy from the early days of thermodynamics, and its application to steam engines and the like. In that context, interest centres on the amount of useful (generally mechanical) work that can be achieved for a given input of heat. As the subject developed, the definition of work was broadened to incorporate *all* types of energy transfer *other than* heat.

work. Although this condition is satisfied by the examples that you will meet in this and other Books in this series, you should always bear in mind that Equation 8.1 is restricted in this way[*].

So far, so good; but how does Equation 8.1 help us to measure enthalpy changes? In particular, how do we determine a value of ΔH for the change implied by Figure 8.1, namely

$$\Delta H = H(T_2, p) - H(T_1, p) = ? \tag{8.2}$$

where $H(T_1, p)$ and $H(T_2, p)$ are enthalpies of our sample of water at the initial and final temperatures (298.15 K and 323.15 K), respectively, and at constant pressure p.

For the purpose of the argument, suppose that you decide to heat your sample *directly* by standing it on an electric hotplate. Suppose further that, like most electrical appliances, the hotplate is marked with its power-rating.

⬤ What is the definition of power? What is its SI unit?

⬤ Power is the *rate* of energy conversion: its SI unit is the watt (W), which is defined as follows: $W = J\,s^{-1}$.

Thus, knowing the power-rating of the hotplate and how long it is switched on, it is a simple matter to calculate the amount of electrical energy converted (= power × time). But there is a snag — a serious one: there is no simple way of telling what proportion of this energy is actually transferred to the system, rather than to the immediate surroundings.

Look back at Equation 8.1. In the case under consideration, no electrical *work* is done on the system (the heating element is not included *in* the system), so $w_{el} = 0$. The crucial problem is that we *cannot* measure the heat q transferred to the system, so we cannot determine the corresponding *enthalpy* change ΔH (Equation 8.2) in this way.

Given that ΔH for a reaction is also defined in terms of heat transfer (Equation 6.1), this is a rather disturbing conclusion. But there is a way around the problem. Our imaginary experiment with a beaker corresponds to one extreme case (and evidently not a particularly useful one) of the general expression in Equation 8.1. What about the other extreme, when $q = 0$?

⬤ How can this be achieved in practice?

⬤ The simplest technique is to thermally insulate — for example, by replacing the beaker in Figure 8.1 with a vacuum flask.

Under these conditions, the only way to transfer energy to the system is to do electrical work on it. The question that now arises is: does this alternative provide a way to evaluate the enthalpy change in Equation 8.2? To examine this issue, try the following Question.

---

[*] To lift this restriction, and hence give a more general statement of the first law, requires the introduction of another, and more fundamental, thermodynamic quantity — the *internal* energy, U, which we shall ignore. For futher comment on the restricted nature of Equation 8.1, see the Appendix, p. 203.

QUESTION 8.1

An electrical heating coil with a power-rating of 100 W is immersed in 100 g of water in a Dewar flask (the scientific equivalent of a vacuum flask). When the current is switched on for 1 minute and 44 seconds, the temperature of the water rises from 298.15 K to 323.15 K.

What is the value of the enthalpy change in Equation 8.2? What assumptions must you make?

So the answer to our question is yes, with the proviso that the system is perfectly insulated. Although an ordinary vacuum flask is far from perfect in this sense, more sophisticated equipment can come close to the ideal limit of no heat transfer. As you saw in Question 8.1, the required value of ΔH can then be determined by simply measuring the electrical work needed to produce the desired temperature rise. In symbols,

$$\Delta H = H(T_2, p) - H(T_1, p) = w_{el} \qquad (8.3)$$

since $q = 0$.

A perfectly insulated system like this is described as being **adiabatically** (from the Greek *adiabatos*, meaning 'impassable') **enclosed**.

We conclude with two further points about the first law, which, for the special cases considered here, takes the form:

$$\Delta H = q + w_{el} \qquad (8.1)$$

First, notice that Equation 8.1 relates explicitly to a *change* in enthalpy. The implication is that only *differences* in enthalpy can be determined, not absolute values. The problem is analogous to that of measuring the heights of mountains and the depths of oceans. In each case, we can, in fact, measure only a *difference* in height. Because sea-level is universally accepted as the reference point for measuring height, however, the reference to 'above' or 'below' sea-level is generally dropped. You will see later how we choose an arbitrary zero for enthalpy.

The second point is that *the change in enthalpy of a system depends only on its initial and final states*, and not at all on how it got from one to the other. The '**state of a system**' simply implies a specification of all the properties necessary to define its condition. For our purposes, listing the temperature, pressure and composition of the system will usually suffice. Thus, in Question 8.1, for example, the initial state of our sample of water (Figure 8.1) could be defined as follows (since the molar mass of water is 18.02 g mol^{-1}): $(100/18.02)$ mol of pure water at a temperature of 298.15 K and a pressure of 1 atm. Likewise, the final state consisted of $(100/18.02)$ mol of pure water at a temperature of 323.15 K and a pressure of 1 atm. In Question 8.1, the change from the initial to the final state was accomplished by putting the sample in an adiabatic enclosure, and performing electrical work on it ($q = 0$; $\Delta H = w_{el}$). However, the same result could have been achieved simply by transferring heat to the system with a hotplate ($w_{el} = 0$; $\Delta H = q$). These two methods are quite different, but because in each case, water is taken from the same initial state to the same final state, ΔH is the same despite the two different pathways. Because of this indifference to the *path* of change, enthalpy is called a **state function**.

8.2 Measuring enthalpy changes: calorimetry

We shall consider two representative examples of enthalpy determinations for reactions — one an endothermic process and the other an exothermic process. In both cases we shall take our cue from the discussion in the previous Section, and concentrate on systems that are, at least approximately, adiabatically enclosed.

The first reaction to be considered is an endothermic process, the vaporization of water at its normal boiling temperature (100 °C (373.15 K) and 1 atm pressure):

$$H_2O(l) = H_2O(g) \tag{8.4}$$

With the background provided by the previous Section, you should be able to work through this example for yourself, so try the following Question.

QUESTION 8.2

A heating coil with a power-rating of 50 W is immersed in water in a Dewar flask, as shown in Figure 8.3: the apparatus is maintained at a constant pressure of 1 atm. When the water boils (at 373.15 K), the amount that vaporizes is measured by condensing it in a weighed flask. With the water boiling in the Dewar flask, 1.62 g of water are condensed over a period of 1 minute and 21 seconds.

(a) What is the enthalpy change associated with the vaporization of 1.62 g of water at its normal boiling temperature?

(b) What is the enthalpy change for the vaporization of 1 mol of water under these conditions?

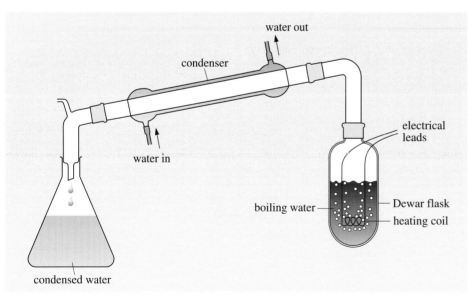

Figure 8.3
Apparatus for measuring an enthalpy of vaporization.

The vaporization or melting of pure substances are examples of **phase transitions**. In principle, the procedure outlined in Question 8.2 can be used to determine ΔH for such transitions in the endothermic direction. In practice, of course, accurate determinations would use equipment, known generally as a **calorimeter**. This is much more sophisticated than a simple Dewar flask, but the essential features of the determination would be the same.

Figure 8.4 Winter scene showing melting ice, and steam issuing from a subterranean hot spring in which boiling temperatures are sometimes reached. The two phase transitions take place at different but constant temperatures.

 Why is the measurement of the *enthalpies of phase transitions* relatively straightforward?

Phase transitions take place at constant temperature (Figure 8.4), so the electrical work done on the system is a direct measure of ΔH at that temperature (provided that $q = 0$, of course!).

Using a thermally insulated calorimeter to measure ΔH for an exothermic *reaction* is not quite as straightforward, as you will now see. The example we have chosen is the neutralization reaction between an acid and a base:

$$H^+(aq) + OH^-(aq) = H_2O(l) \tag{8.5}$$

In a typical experiment, 0.1 litre of HCl(aq) and 0.1 litre of NaOH(aq), both of concentration 2.0 mol litre^{-1} and both at 300 K, were mixed in a thermally insulated calorimeter, at a constant pressure of 1 atm.

If the reaction is exothermic, what do you predict will happen?

Since the energy released by the reaction cannot be transferred to the surroundings, the temperature should rise.

It does: to 311 K in the experiment mentioned above. This change is illustrated schematically in Figure 8.5, where we have used the generalized notation of 'reactants' for $H^+(aq) + OH^-(aq)$ and 'product' for $H_2O(l)$.

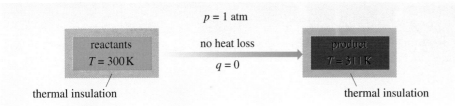

But how do we relate this *temperature* change to the required value of ΔH at a fixed temperature of 300 K? To examine this question, let us call reactants at the initial temperature (300 K) 'state 1', and product at the final temperature (311 K) 'state 2'. Then the enthalpy change for the process in Figure 8.5 can be represented as follows:

$$\Delta H_{1\rightarrow2} = H_2 - H_1$$
$$= H(\text{product, 311 K}) - H(\text{reactants, 300 K}) \qquad (8.6)$$

To determine the value of $\Delta H_{1\rightarrow2}$ requires our statement of the first law for a change at constant pressure, Equation 8.1:

$$\Delta H = q + w_{el} \qquad (8.1)$$

Now, in the experiment described earlier, no electrical work is done on the system, so $w_{el} = 0$. Moreover, if we assume that the calorimeter is perfectly insulated (admittedly an idealization), then the heat transfer will also be zero. Combining these results,

$$w_{el} = 0 \text{ and } q = 0$$

so

$$\Delta H_{1\rightarrow2} = 0$$

A surprising and somewhat alarming conclusion at first sight! It becomes less so once we recognize that the change illustrated in Figure 8.5 is not the one we set out to measure. Indeed, for any experiment like this, $\Delta H_{1\rightarrow2}$ will always be, at least approximately, zero.

● What change are we interested in?

● Reactants at the initial temperature going to product at the *same* temperature.

If we call *product* at the initial temperature (300 K) 'state 3', then the enthalpy change we are after is

$$\Delta H_{1\rightarrow3} = H(\text{product, 300 K}) - H(\text{reactants, 300 K}) \qquad (8.7)$$

The two enthalpy changes that we have described are illustrated schematically in Figure 8.6. But how can we relate the result of our first experiment to the desired enthalpy change, $\Delta H_{1\rightarrow3}$?

● Remembering that H is a state function, use Figure 8.6 to derive an expression for $\Delta H_{1\rightarrow3}$ in terms of $\Delta H_{1\rightarrow2}$ and the enthalpy change between states 2 and 3.

● As H is a state function, then, with reference to Figure 8.6, we can get from state 1 to state 2 either directly (as in the experiment described earlier) or by way of state 3. In other words, you can complete the cycle in Figure 8.6 by drawing in an arrow *from* state 3 to state 2 (see Figure 8.7): in symbols,

$$\Delta H_{1\rightarrow2} = \Delta H_{1\rightarrow3} + \Delta H_{3\rightarrow2}$$

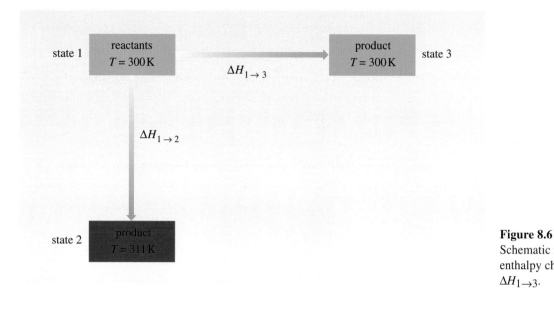

Figure 8.6
Schematic representation of the enthalpy changes $\Delta H_{1\rightarrow2}$ and $\Delta H_{1\rightarrow3}$.

so

$$\Delta H_{1\rightarrow3} = \Delta H_{1\rightarrow2} - \Delta H_{3\rightarrow2} \tag{8.8}$$

where

$$\Delta H_{3\rightarrow2} = H(\text{product}, 311\text{ K}) - H(\text{product}, 300\text{ K}) \tag{8.9}$$

Now concentrate on Equations 8.8 and 8.9. You have just seen that $\Delta H_{1\rightarrow2}$ is effectively zero: but how can $\Delta H_{3\rightarrow2}$ be determined? The answer is that a second experiment is required: it's usually called a *calibration* experiment. For example, to complete our determination, the product (water, or, more accurately, an aqueous solution of sodium chloride, in this case) is first allowed to cool back to the original temperature, 300 K. When a heating coil with a power-rating of 50 W is immersed in the liquid, and the current is switched on for 220 s, the temperature rises back to 311 K.

🔘 What is the value of $\Delta H_{3\rightarrow2}$ (Equation 8.9) in this case?

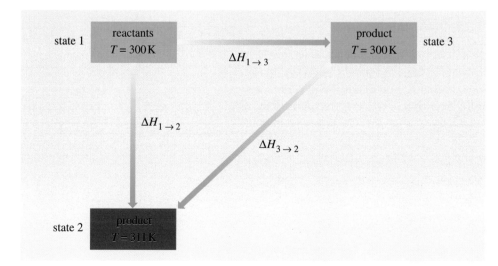

Figure 8.7
The cycle of Figure 8.6 completed by the addition of an arrow from state 3 to state 2.

● Assuming once again that the system is perfectly insulated, $q = 0$, so

$$\Delta H_{3\rightarrow2} = w_{el} = (50\,J\,s^{-1}) \times 220\,s$$
$$= 11\,000\,J$$
$$= 11.0\,kJ$$

● What is the measured value of ΔH for the neutralization reaction in Equation 8.5 at a temperature of 300 K and a pressure of 1 atm?

● From Equation 8.8, the required enthalpy change is given by:

$$\Delta H_{1\rightarrow3} = \Delta H_{1\rightarrow2} - \Delta H_{3\rightarrow2}$$
$$= 0 - 11\,kJ$$
$$= -11\,kJ$$

Notice that ΔH for this process is, as expected, a negative quantity: the reaction *is* exothermic.

To summarize, the essential elements of this sort of determination are:

1   To perform the reaction under conditions in which the heat flow to or from the system is kept as close to zero as possible: the *temperature* change resulting from the reaction is measured.

2   To relate this temperature change to the desired enthalpy change: this is achieved by means of a second, calibration, experiment in which a measured amount of electrical energy is transferred to the reaction *product(s)*.

This completes our brief discussion of the experimental measurement of enthalpy changes. In practice, many different sorts of calorimeter have been developed for studying different types of reaction (Figure 8.8). Nevertheless, the principles underlying their design and use can always be traced back to the first law of thermodynamics, as for the examples considered here.

Figure 8.8
A simple research calorimeter, showing the Dewar flask used as the reaction vessel. The plastic disc covers the Dewar during measurements, and contains apertures for the stirring facilities (note the pulley wheel) and electrical leads to an internal heating coil and thermistor (a temperature-measuring device). At the top of the frame is the motor that drives the stirrer.

8.3 Molar enthalpy changes

As measured in a calorimeter, the value of ΔH for a reaction depends on the amounts of reactants used in the experiment. To have to state the amounts actually chosen, in addition to the temperature and pressure, would hopelessly complicate the comparison of ΔH values from different sources. Thus, experimental results are usually quoted as *molar enthalpy changes*, and denoted by attaching a subscript to the symbol ΔH, as ΔH_m.

You met one example in the answer and comment to Question 8.2, namely the molar enthalpy of vaporization of water. As you saw, this represents the molar enthalpy change for the process in Equation 8.4:

$$H_2O(l) = H_2O(g) \tag{8.4}$$

where $\Delta H_m = 44.3\,kJ\,mol^{-1}$.

But how about a more complicated reaction, such as the one between sodium and chlorine that you met in Section 5?

$$Na(s) + \frac{1}{2}Cl_2(g) = NaCl(s) \qquad\qquad (5.2)$$

In this case, the molar enthalpy change is defined as the enthalpy change when one mole of sodium reacts with half a mole of $Cl_2(g)$ to give one mole of sodium chloride. Now suppose that, in a particular experiment, 2.3 g (0.1 mol) of sodium reacts completely with chlorine, and that the corresponding enthalpy change is found to be $\Delta H = -41.12$ kJ.

⬤ What is the molar enthalpy change for the reaction in Equation 5.2?

⬤ When 0.1 mol of sodium reacts completely with chlorine, $\Delta H = -41.12$ kJ. But we require ΔH for the case when 1 mol of sodium reacts completely with chlorine. Thus,

$$\Delta H_m = \frac{-41.12 \text{ kJ}}{0.1 \text{ mol}} = -411.2 \text{ kJ mol}^{-1}$$

If you look back at Question 6.1, you will see that this corresponds to the value of ΔH quoted for the reaction in Equation 5.2. Indeed, all of the enthalpy changes cited in Sections 6 and 7 are actually values of ΔH_m, and should strictly have the unit kJ mol^{-1}.

But there is a more important point about the value of ΔH_m: *it depends on which balanced equation you choose to represent the reaction*. The molar enthalpy change of an equation, such as Equation 5.2, is the enthalpy change of the reaction involving a certain number of moles of each substance in the equation. For each substance, that number is the one preceding its formula in the equation. Thus, if we multiply Equation 5.2 through by four, then the value of ΔH_m must also be quadrupled:

$$Na(s) + \frac{1}{2}Cl_2(g) = NaCl(s); \Delta H_m = -411.2 \text{ kJ mol}^{-1} \qquad (5.2)$$
$$4Na(s) + 2Cl_2(g) = 4NaCl(s); \Delta H_m = -1\,644.8 \text{ kJ mol}^{-1} \qquad (8.10)$$

This is because the molar enthalpy change of Equation 8.10 refers to the enthalpy change of the reaction in which 4 moles of sodium react with 2 moles of chlorine to give 4 moles of sodium chloride. So, when we quote ΔH_m values in kJ mol^{-1}, the 'per mole' (mol^{-1}) has a special meaning; it is 'per mole' of a chosen, balanced equation, not 'per mole' of any particular substance. *Consequently, it is meaningful to quote a value of ΔH_m, for a reaction only if you know which balanced equation you are talking about.*

QUESTION 8.3

Use the experimental results quoted in Section 8.2 to calculate ΔH_m for the following reaction at 300 K and 1 atm pressure:

$$H^+(aq) + OH^-(aq) = H_2O(l) \qquad\qquad (8.5)$$

QUESTION 8.4

A calorimeter is used to determine the enthalpy change when 0.50 mol of lithium reacts completely with oxygen to give 0.25 mol of lithium oxide, Li_2O. The value obtained is $\Delta H = -149.5$ kJ. Calculate and write down the values of ΔH_m for the following reactions:

$$2Li(s) + \tfrac{1}{2}O_2(g) = Li_2O(s)$$
$$4Li(s) + O_2(g) = 2Li_2O(s)$$
$$8Li(s) + 2O_2(g) = 4Li_2O(s)$$

8.4 Summary of Section 8

1   Energy can be transferred to a system either by heating it or by doing electrical work on it. For a change at constant pressure, our statement of the first law of thermodynamics expresses this result in terms of the enthalpy of the system:

$$\Delta H = q + w_{el}$$

2   Only differences in enthalpy can be determined, never absolute values.

3   Enthalpy is a state function: its value depends only on the state of the system, as specified by its temperature, pressure and composition.

4   For an adiabatically enclosed system ($q = 0$), our statement of the first law provides a means of measuring:

(i)   the enthalpy increment, $\Delta H = H(T_2) - H(T_1)$, for a pure substance, or mixture of substances, at constant pressure;

(ii) the enthalpy of transition of a pure substance, at constant temperature and pressure;

(iii) when combined with a suitable calibration experiment, the enthalpy change for an exothermic or endothermic reaction at constant temperature and pressure.

Now try the following Questions.

QUESTION 8.5

A sample of 1.308 g of zinc reacted with an excess of dilute hydrochloric acid in a thermally insulated calorimeter. The temperature of the calorimeter and its contents rose from 298.15 K to 299.5 K. The calorimeter and its contents were then cooled back to 298.15 K. A heating coil, with a power rating of 50 W in the calorimeter, was switched on for 61.5 s in order to raise the temperature back to 299.5 K. You may assume that the pressure remained constant throughout. Take the mass of one mole of zinc as 65.38 g.

Calculate the value of ΔH_m at 298.15 K for the reactions:

(i) $Zn(s) + 2H^+(aq) = Zn^{2+}(aq) + H_2(g)$

(ii) $\tfrac{1}{2}Zn(s) + H^+(aq) = \tfrac{1}{2}Zn^{2+}(aq) + \tfrac{1}{2}H_2(g)$

In answering this question, give an indication of any assumptions you make.

QUESTION 8.6

The molar enthalpy change for the reaction between aluminium and chlorine is -704.2 kJ mol^{-1}. What is deficient about this statement?

ENTHALPIES OF REACTION: A DATABASE

9

If a reaction happens quickly, its enthalpy change can be determined in a calorimeter. But many reactions occur only very slowly, or not at all; how then can we determine their enthalpy changes? Consider first the decomposition of sodium chloride:

$$NaCl(s) = Na(s) + \tfrac{1}{2}Cl_2(g) \tag{5.1}$$

As noted in Section 5, equilibrium lies to the left, and the reaction does not occur.

⬤ How can its enthalpy change be obtained?

⬤ The answer is to determine the enthalpy change of the reverse reaction, which happens quickly:

$$Na(s) + \tfrac{1}{2}Cl_2(g) = NaCl(s); \Delta H_m = -411.2 \, kJ \, mol^{-1} \tag{5.2}$$

The enthalpy change of Reaction 5.1 is minus that of Reaction 5.2; that is,

$$\Delta H_m(5.1) = +411.2 \, kJ \, mol^{-1}$$

Here we have used the fact that enthalpy is a state function. In Equation 5.1, sodium chloride breaks down into sodium and chlorine; in Equation 5.2, the sodium and chlorine recombine to regenerate sodium chloride. Provided that the sodium chloride that we start with is at the same temperature and pressure as the sodium chloride that we finish with, the overall enthalpy change must have been zero. Thus,

$$\Delta H_m(5.1) + \Delta H_m(5.2) = 0$$
$$\Delta H_m(5.1) = -\Delta H_m(5.2) = 411.2 \, kJ \, mol^{-1} \tag{9.1}$$

Unfortunately, the procedure that we have just followed cannot always be used, because of a lack of data. For example, in Activity 2.1, you saw that the reaction between copper and hydrogen ions does not happen:

$$Cu(s) + 2H^+(aq) = Cu^{2+}(aq) + H_2(g) \tag{2.9}$$

But if one tries to determine the enthalpy change of this reaction by studying the reverse reaction between hydrogen gas and $Cu^{2+}(aq)$, one finds that this reaction does not occur either. Equilibrium must lie either to the left or the right of Equation 2.9, but the movement towards it must be a very slow process. How then can $\Delta H_m(2.9)$ be determined?

9.1 Hess's law

Consider the reactions of metallic zinc, firstly with aqueous hydrogen ions and secondly with copper ions, $Cu^{2+}(aq)$. You saw both reactions in Activities 2.1 and 2.2, and both were reasonably fast. Suppose then that we determine the enthalpy of each reaction in a calorimeter. The results are:

$$Zn(s) + 2H^+(aq) = Zn^{2+}(aq) + H_2(g); \Delta H_m = -153.9 \, kJ \, mol^{-1} \tag{9.2}$$
$$Zn(s) + Cu^{2+}(aq) = Zn^{2+}(aq) + Cu(s); \Delta H_m = -218.7 \, kJ \, mol^{-1} \tag{9.3}$$

Now chemical changes, along with their enthalpy changes, can be added and subtracted. This technique was used in the previous Section.

● Subtract Equation 9.3 from Equation 9.2. What reaction does the resulting equation represent, and what is its enthalpy change?

● On subtraction, zinc and its ions disappear:

$$2H^+(aq) - Cu^{2+}(aq) = H_2(g) - Cu(s); \Delta H_m = 64.8 \text{ kJ mol}^{-1} \qquad (9.4)$$

Carrying the formulae with minus signs in front of them over to the other side of the equation, we get

$$Cu(s) + 2H^+(aq) = Cu^{2+}(aq) + H_2(g); \Delta H_m = 64.8 \text{ kJ mol}^{-1} \qquad (9.5)$$

The result is Equation 2.9, along with the enthalpy change that we were looking for (Figure 9.1). The algebraic manipulation of equations and their enthalpy changes is an application of **Hess's law**. This states that:

> An energy change for a chemical reaction is the same whether the reaction takes place in just one step, or by a number of separate steps whose sum is equal to the one-step process.

Like the procedure of Section 9, Hess's law depends on the fact that enthalpy is a state function, a point that will become clearer in Section 9.3.1. Here, you should particularly note that a database consisting of, say, 100 selected reactions and their enthalpy changes is very fertile, because the manipulation procedure of this Section allows us to calculate the enthalpy changes of many more than 100 reactions from it. We now turn to the question of what type of reaction to put into the database.

Figure 9.1
Copper does not react with dilute hydrochloric acid: in particular, Reaction 9.5 does not happen. Nevertheless, Hess's law tells us what its enthalpy change would be if it did.

9.2 Standard enthalpy changes

When, in the last Section, you subtracted Equation 9.3 from Equation 9.2, the zinc metal cancelled out. But this is legitimate only if the zinc in Equation 9.2 is in the same form as the zinc in Equation 9.3. So if our database of enthalpies of reaction is to be useful, a particular reactant or product must always appear in the same state. We ensure that this is so by specifying its pressure, composition and temperature.

Of these, pressure is the least important, because the dependence of enthalpies of reaction on pressure is only slight. All reactants and products are specified as being at a standard pressure of 100 kPa. This is only slightly less than the definition of atmospheric pressure (1 atm ≡ 101.325 kPa). Next we take composition. For distinct substances, such as the zinc metal and hydrogen gas in Equation 9.2, or the liquid water in Equation 8.4, we specify that the substance must be pure. For dissolved substances, such as the ions $H^+(aq)$ and $Zn^{2+}(aq)$ in Equation 9.2, some kind of concentration must be specified as well. This is a complicated business for reasons that we shall not go into; all we note here is that it results in ΔH_m values that are appropriate for ions in dilute solutions.

This combined specification of pressure and composition gives us what is called the 'standard state' of a substance. For example, the standard state of zinc metal is the pure metal at a pressure of 100 kPa. When reactants and products are in these standard states, the molar enthalpy of reaction is labelled with a superscript $\ominus$, thus: $\Delta H_m^{\ominus}$.

This quantity, $\Delta H_m^\ominus$, is called the **standard molar enthalpy change** *. The values of $\Delta H_m^\ominus$ now depend only on the temperature. Most compilations, including that in the *Data Book* (on the CD-ROM), list values of $\Delta H_m^\ominus$ at a temperature of 298.15 K (25 °C). All values of $\Delta H_m^\ominus$ quoted so far in this Book, have, in fact, been $\Delta H_m^\ominus$ values at 298.15 K.

9.3 Standard enthalpies of formation

What kinds of standard enthalpy change make up our database? One enthalpy change is chosen for each chemical species that is included. It is called the standard enthalpy of formation. You will be dealing with two kinds of standard enthalpy of formation: those of pure substances, and those of aqueous ions.

9.3.1 Enthalpies of formation of pure substances

The **standard enthalpy of formation** of a pure substance, $\Delta H_f^\ominus$ at 298.15 K, is defined as the standard molar enthalpy change at 298.15 K of the reaction in which one formula unit of the substance is formed from its elements in their reference states.

To clarify this definition, we shall look at the examples in Table 9.1, which have been taken from the *Data Book*. For liquid water, Table 9.1 tells us that $\Delta H_f^\ominus$ (H_2O, l) $= -285.8$ kJ mol^{-1}. One formula unit is H_2O(l), which tells us that the *balanced* formation reaction (Figure 9.2) is:

$$H_2(g) + \tfrac{1}{2}O_2(g) = H_2O(l) \tag{9.6}$$

and its standard enthalpy change is $\Delta H_m^\ominus = -285.8$ kJ mol^{-1}.

Notice that the physical state of the pure substance, *liquid* water, must be specified explicitly. This explains why there is a second entry for the formula H_2O in Table 9.1.

- Write down the equation to which this second entry refers, along with its standard enthalpy change.

- The second entry is for *gaseous* water, H_2O(g):

$$H_2(g) + \tfrac{1}{2}O_2(g) = H_2O(g); \quad \Delta H_m^\ominus = -241.8 \text{ kJ mol}^{-1} \tag{9.7}$$

Our definition referred to formation from the *reference* states of the elements. These are nearly always the commonest form in which the element exists at 298.15 K and 100 kPa — effectively room temperature and pressure. Thus, the diatomic gas is the reference state for hydrogen, oxygen and chlorine, etc., and the liquid is the reference state for mercury and bromine. Some elements exist in more than one crystalline form as solids at room temperature and pressure. The commonest form is then usually chosen as the reference state. For example, graphite rather than diamond is the chosen reference state for carbon.

The reference states of the elements can easily be identified in data tables because their $\Delta H_f^\ominus$ values must be zero.

* $\Delta H_m^\ominus$ is pronounced delta-h-m-standard.

Table 9.1
Standard enthalpies of formation of some pure substances at 298.151K

Table 9.1
Standard enthalpies of formation of some pure substances at 298.151K

Substance	$\Delta H_f^\ominus$/kJ mol^{-1}
H_2O(l)	−285.8
H_2O(g)	−241.8
H_2(g)	0
O_2(g)	0
I_2(s)	0
I_2(g)	62.4
Al(s)	0
Al_2O_3(s)	−1 675.7
Ca(s)	0
CaO(s)	−635.1
$CaCO_3$(s)	−1 206.9
C(graphite)	0
C(diamond)	1.9
CO_2(g)	−393.5
N_2(g)	0
NO_2(g)	33.2
N_2O_4(g)	9.2

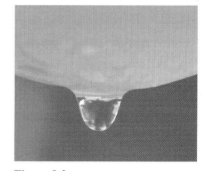

Figure 9.2
Condensed water vapour on a solid surface over which a jet of burning hydrogen gas has been played. The hydrogen flame is coloured red by elements in the gas jet.

⬤ Why is this?

⬤ Using $H_2(g)$ as an example, by definition, $\Delta H_f^{\ominus}(H_2, g)$ is $\Delta H_m^{\ominus}$ for the reaction in which $H_2(g)$ is formed from hydrogen in its reference state. But this reference state *is* $H_2(g)$, so the formation reaction is

$$H_2(g) = H_2(g) \tag{9.8}$$

This reaction involves no change whatsoever, so $\Delta H_m^{\ominus}$, and therefore $\Delta H_f^{\ominus}(H_2, g)$ is zero.

⬤ To what reaction does the entry $\Delta H_f^{\ominus}(I_2, g) = 62.4\,kJ\,mol^{-1}$ refer?

⬤ Iodine is a solid at room temperature and pressure so, as the zero entry in Table 9.1 shows, $I_2(s)$ is the reference state. $\Delta H_f^{\ominus}(I_2, g)$ refers to the reaction (Figure 9.3):

$$I_2(s) = I_2(g);\ \Delta H_m^{\ominus} = 62.4\,kJ\,mol^{-1} \tag{9.9}$$

Figure 9.3
The modest enthalpy change of Reaction 9.9 shows that the forces that bind one iodine molecule to another in solid iodine are not strong. In a sealed gas jar at room temperature, the purple colour of iodine vapour is visible above small beads of solid iodine.

QUESTION 9.1

Write down the reactions to which $\Delta H_f^{\ominus}(Al_2O_3, s)$ and $\Delta H_f^{\ominus}(CaCO_3, s)$ refer. Use Table 9.1 to state their standard enthalpy changes.

9.3.2 Calculating standard enthalpies of reaction

How do we use our database of values to calculate the standard enthalpy change of any chemical reaction? Let's do it for the decomposition of calcium carbonate (Figure 9.4):

$$CaCO_3(s) = CaO(s) + CO_2(g) \tag{9.10}$$

Table 9.1 gives the $\Delta H_f^{\ominus}$ value for each substance in the equation.

⬤ Write equations, with enthalpy changes, to which these $\Delta H_f^{\ominus}$ values refer.

⬤ See the following table.

Reaction	$\Delta H_m^{\ominus}$ /kJ mol^{-1}	
$Ca(s) + \frac{3}{2}O_2(g) + C(graphite) = CaCO_3(s)$	$-1\,206.9$	(9.11)
$Ca(s) + \frac{1}{2}O_2(g) = CaO(s)$	-635.1	(9.12)
$C(graphite) + O_2(g) = CO_2(g)$	-393.5	(9.13)

Figure 9.4
A methane–oxygen torch is used to decompose limestone (calcium carbonate). Quicklime (calcium oxide, CaO) is produced, and at these very high temperatures becomes incandescent, emitting an intense white light. This was the 'limelight' used in nineteeth-century theatres.

We now use the technique of Section 9.1: we perform an addition and/or subtraction of these equations and their $\Delta H_m^\ominus$ values, which leaves us with Equation 9.10 and its $\Delta H_m^\ominus$ value. (This is the equivalent of applying Hess's law.) As the following table shows, this can be done by just reversing Equation 9.11 and then adding the three together:

Reaction	$\Delta H_m^\ominus$ /kJ mol^{-1}	
$CaCO_3(s) = Ca(s) + \frac{3}{2}O_2(g) + C(graphite)$	$-(-1\,206.9)$	(9.14)
$Ca(s) + \frac{1}{2}O_2(g) = CaO(s)$	-635.1	(9.12)
$C(graphite) + O_2(g) = CO_2(g)$	-393.5	(9.13)
$CaCO_3(s) = CaO(s) + CO_2(g)$	$+178.3$	(9.10)

Although it is always possible to write down a table like this, there is a simpler and more general way. What the table does is to add the formation reactions of $CaO(s)$ and $CO_2(g)$, and then subtract the formation reaction of $CaCO_3(s)$. It also does the same thing for the $\Delta H_f^\ominus$ values. Overall then, it adds the $\Delta H_f^\ominus$ values for the *products* of Equation 9.10, and substracts the $\Delta H_f^\ominus$ value for the *reactant*:

$$\Delta H_m^\ominus = \Delta H_f^\ominus(CaO, s) + \Delta H_f^\ominus(CO_2, g) - \Delta H_f^\ominus(CaCO_3, s) \qquad (9.15)$$

This is a completely general result: it can be used to calculate $\Delta H_m^\ominus$ for *any* chemical reaction. Thus, for the general reaction,

$$a\text{A} + b\text{B} + \ldots = x\text{X} + y\text{Y} + \ldots \qquad (9.16)$$

(where the lower case a, b, x, y, … etc, are the numbers that precede the formulae — also known as 'coefficients' — in the balanced chemical equation), we can write:

$$\Delta H_m^\ominus = \{x\,\Delta H_f^\ominus(\text{X}) + y\,\Delta H_f^\ominus(\text{Y}) + \ldots\} - \{a\,\Delta H_f^\ominus(\text{A}) + b\,\Delta H_f^\ominus(\text{B}) + \ldots\} \qquad (9.17)$$

or, making use of the symbol Σ (capital Greek sigma) to represent the act of adding things together,

$$\Delta H_m^\ominus = \Sigma\,\Delta H_f^\ominus(\text{products}) - \Sigma\,\Delta H_f^\ominus(\text{reactants}) \qquad (9.18)$$

When using this expression, you should always remember that each value of $\Delta H_f^\ominus$ must be multiplied by the number that precedes its formula in the balanced chemical equation.

A further example should make this clearer.

Use information from Table 9.1 to calculate $\Delta H_m^\ominus$ at 298.15 K for the reaction
$$N_2O_4(g) = 2NO_2(g) \qquad (9.19)$$

From Equation 9.18, and noting that the value for the product, NO_2, must be multiplied by two:

$$\Delta H_m^\ominus = 2\,\Delta H_f^\ominus(NO_2, g) - \Delta H_f^\ominus(N_2O_4, g) \qquad (9.20)$$

$$= \{2 \times 33.2 - (9.2)\} = 57.2\,\text{kJ mol}^{-1}$$

If you would like some more practice using Equation 9.18 at this stage, then try the following Question.

QUESTION 9.2

Use information from the *Data Book* to calculate $\Delta H_m^{\ominus}$ at 298.15 K for the following reactions:

(i) $Na(s) + \frac{1}{2}Cl_2(g) = NaCl(s)$ (5.2)

(ii) $PbO(s) + C(graphite) = Pb(s) + CO(g)$ (3.1)

(iii) $SOCl_2(l) + H_2O(l) = SO_2(g) + 2HCl(g)$ (6.4)

(iv) $3K(s) + AlCl_3(s) = 3KCl(s) + Al(s)$ (3.9)

9.3.3 Standard enthalpies of formation of aqueous ions

Unlike the examples in Question 9.2, the reactions that you observed in Activities 2.1 and 2.2 involved aqueous ions as well as neutral species. To extend the procedure represented by Equation 9.18 to reactions of this type requires a definition of $\Delta H_f^{\ominus}$ for an aqueous ion. However, because ions are charged, this presents problems.

Consider the chloride ion, $Cl^-(aq)$. Our definition of $\Delta H_f^{\ominus}$ for pure substances suggests that one possible formation reaction is the formation of the ion from chlorine gas (the reference state of the element) with an extra electron to create the negative charge:

$$\frac{1}{2}Cl_2(g) + e^- = Cl^-(aq)^*$$ (9.21)

Unfortunately, reactions of this kind involve only the formation of a single type of aqueous ion. They cannot be performed in a laboratory without forming ions of opposite charge at the same time, so $\Delta H_m^{\ominus}(9.21)$ is not experimentally measurable. The best we can do is to study reactions in which *two* kinds of ion are formed from their constituent elements; for example

$$\frac{1}{2}H_2(g) + \frac{1}{2}Cl_2(g) = H^+(aq) + Cl^-(aq)$$ (9.22)

We can, for example, determine $\Delta H_m^{\ominus}$ for this reaction by combining hydrogen and chlorine gases (Figure 9.5) in the presence of water. The gases form hydrogen chloride, $HCl(g)$; this then dissolves in the water to give hydrochloric acid, which contains $H^+(aq)$ and $Cl^-(aq)$. The value of $\Delta H_m^{\ominus}(9.22)$ is found to be $-167.2\ kJ\ mol^{-1}$.

⬤   What relationship can reasonably be assumed to exist between $\Delta H_m^{\ominus}(9.22)$, and $\Delta H_f^{\ominus}(H^+, aq)$ and $\Delta H_f^{\ominus}(Cl^-, aq)$?

⬤   Since, in Equation 9.22, $H^+(aq)$ and $Cl^-(aq)$ are formed together from the reference states of their elements, it seems reasonable to write

$$\Delta H_m^{\ominus}(9.22) = \Delta H_f^{\ominus}(H^+, aq) + \Delta H_f^{\ominus}(Cl^-, aq)$$ (9.23)

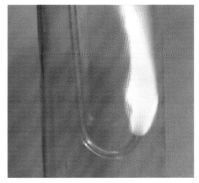

Figure 9.5
A glass jet of hydrogen burning in a jar of chlorine to give hydrogen chloride, HCl. In the presence of water, the HCl gas dissolves to give aqueous hydrogen and chloride ions.

---

* We again use (aq) to indicate that the ion is in aqueous solution, but you may be puzzled by the absence of any reference to water on the left-hand side of Equation 9.21. Although one could write 'excess water' or '(aq)' on the left-hand side of the equations, it is *conventional* not to do so. The same is true for equations that you are more familiar with, like Equation 6.3.

The problem now becomes one of assigning one part of $\Delta H_m^{\ominus}(9.22)$ to $\Delta H_f^{\ominus}(\text{H}^+, \text{aq})$, and the rest to $\Delta H_f^{\ominus}(\text{Cl}^-, \text{aq})$. To get around this problem, an arbitrary convention is introduced: we simply decide that

$$\Delta H_f^{\ominus}(\text{H}^+, \text{aq}) = 0 \qquad\qquad (9.24)$$

● Given this convention, and the fact that $\Delta H_m^{\ominus}(9.22) = -167.2\,\text{kJ mol}^{-1}$, what is $\Delta H_f^{\ominus}(\text{Cl}^-, \text{aq})$?

● Substitution of Equation 9.24 into Equation 9.23 gives the value $\Delta H_f^{\ominus}(\text{Cl}^-, \text{aq}) = -167.2\,\text{kJ mol}^{-1}$.

By using this convention and the all-important Equation 9.18, we can build up a collection of $\Delta H_f^{\ominus}$ values for individual aqueous ions. For example, we can now determine $\Delta H_f^{\ominus}(\text{Na}^+, \text{aq})$ from the experimental value of the standard molar enthalpy of solution of sodium choride, which was given in Section 6:

$$\text{NaCl(s)} = \text{Na}^+(\text{aq}) + \text{Cl}^-(\text{aq}); \; \Delta H_m^{\ominus} = 3.9\,\text{kJ mol}^{-1} \qquad\qquad (6.3)$$

● We have just set $\Delta H_f^{\ominus}(\text{Cl}^-, \text{aq}) = -167.2\,\text{kJ mol}^{-1}$, and we know that $\Delta H_f^{\ominus}(\text{NaCl}, \text{s}) = -411.2\,\text{kJ mol}^{-1}$. What is $\Delta H_f^{\ominus}(\text{Na}^+, \text{aq})$?

● Applying Equation 9.18 to Equation 6.3,

$$\Delta H_m^{\ominus}(6.3) = \Delta H_f^{\ominus}(\text{Na}^+, \text{aq}) + \Delta H_f^{\ominus}(\text{Cl}^-, \text{aq}) - \Delta H_f^{\ominus}(\text{NaCl}, \text{s}) \qquad (9.25)$$

so

$$\begin{aligned}
\Delta H_f^{\ominus}(\text{Na}^+, \text{aq}) &= \Delta H_m^{\ominus}(6.3) - \Delta H_f^{\ominus}(\text{Cl}^-, \text{aq}) + \Delta H_f^{\ominus}(\text{NaCl}, \text{s}) \\
&= \{+3.9 - (-167.2) + (-411.2)\}\,\text{kJ mol}^{-1} \\
&= -240.1\,\text{kJ mol}^{-1}
\end{aligned}$$

If you check your *Data Book*, you will find that this is indeed the value of $\Delta H_f^{\ominus}(\text{Na}^+, \text{aq})$. Moreover, by exploiting Equation 9.18, you can now use the database of standard enthapies of formation to calculate the standard enthalpy changes of a host of chemical reactions involving both pure substances and aqueous ions.

Finally, it is not difficult to show that the assignment made in Equation 9.24 is equivalent to defining $\Delta H_f^{\ominus}$ for a possible aqueous ion, such as $\text{Na}^+(\text{aq})$, as $\Delta H_m^{\ominus}$ for a reaction of the type:

$$\text{Na(s)} + \text{H}^+(\text{aq}) = \text{Na}^+(\text{aq}) + \tfrac{1}{2}\text{H}_2(\text{g}) \qquad\qquad (9.26)$$

Similar formation reactions can be written for other aqueous ions, but we defer further discussion of this until Section 11.2.

9.4 Summary of Section 9

1   Standard enthalpies of reaction are molar enthalpy changes of reactions that take place at constant temperature, and in which reactants and products are in their standard states. For distinct substances, these standard states are the pure substance at a pressure of 100 kPa (0.987 atm). The constant temperature is usually defined as 298.15 K.

2   The standard enthalpy of formation of a pure substance is the standard molar enthalpy change of the reaction in which one formula unit of the substance is formed from its elements in their reference states at a constant temperature (usually 298.15 K).

3   The standard molar enthalpy changes of reactions can be calculated from a database of standard enthalpies of formation by applying Hess's law using the equation

$$\Delta H_m^{\ominus} = \Sigma \Delta H_f^{\ominus}(\text{products}) - \Sigma \Delta H_f^{\ominus}(\text{reactants}) \tag{9.18}$$

4   By using the convention $\Delta H_f^{\ominus}(\text{H}^+, \text{aq}) = 0$, standard enthalpies of formation of aqueous ions can be included in the database. Equation 9.18 then provides the standard enthalpies of reactions that include ions.

QUESTION 9.3

Using information from the *Data Book*, calculate $\Delta H_m^{\ominus}$ at 298.15 K for the following reactions:

(i) $\text{Zn(s)} + 2\text{H}^+(\text{aq}) = \text{Zn}^{2+}(\text{aq}) + \text{H}_2(\text{g})$ (2.4)

(ii) $\text{Mg(s)} + 2\text{Ag}^+(\text{aq}) = \text{Mg}^{2+}(\text{aq}) + 2\text{Ag(s)}$ (2.13)

(iii) $\text{H}^+(\text{aq}) + \text{OH}^-(\text{aq}) = \text{H}_2\text{O(l)}$ (8.5)

QUESTION 9.4

Like sodium, magnesium burns in chlorine:

$$\text{Mg(s)} + \text{Cl}_2(\text{g}) = \text{MgCl}_2(\text{s}); \quad \Delta H_m^{\ominus}(298.15\,\text{K}) = -641.3\,\text{kJ mol}^{-1} \tag{9.27}$$

The standard molar enthalpy of solution of $\text{MgCl}_2(\text{s})$ is $-160.0\,\text{kJ mol}^{-1}$, and $\Delta H_f^{\ominus}(\text{Cl}^-, \text{aq})$ is $-167.2\,\text{kJ mol}^{-1}$. What is the standard enthalpy of formation of $\text{Mg}^{2+}(\text{aq})$?

ENTROPY CHANGES

10

At this stage, a reminder of where we stand may be useful. In Section 7, we used the second law of thermodynamics to find a quantity, $\Delta H - T\Delta S$, which does for chemical reactions at constant temperature and pressure, what potential energy does for the motion of a stone in a gravitational field. The quantity consists of two terms. We then set out to show how each of them could be experimentally determined. In Sections 8 and 9, you were shown how the first term, ΔH, could be measured, and how the measurements could be used to build up a database from which other ΔH values could be calculated. It now remains to repeat the process for the second term, $-T\Delta S$. The crucial task here is the determination of the entropy change in a reaction, ΔS, and it is to this that we now turn.

10.1 Determining entropy changes

To find out how to determine entropies of reaction, we return to the definitions given in Section 7.1. For this purpose, cases 1 and 3 are the important ones:

1   When heat is transferred to or from a pure substance at constant temperature and pressure, the change in the entropy of the substance is

$$\Delta S_{\text{substance}} = \frac{q}{T} \qquad (7.2)$$

3   When a chemical reaction takes place at constant temperature and pressure, the change in entropy associated with the reaction is given by

$$\Delta S_{\text{reac.}} = S(\text{products}) - S(\text{reactants}) \qquad (7.4)$$

Before going further, notice that the definition in Equation 7.4 implies that entropy, like enthalpy (compare Equation 6.2, p. 42) is a state function. In other words, the *change* in entropy of a system depends only on its initial and final states, and not at all on how it proceeded from one to the other.

Now, as we said in Section 7.1, the use of Equation 7.4 depends on the determination of the entropies of pure substances via the expression in Equation 7.2. So we shall start with the entropies of pure substances.

10.2 Entropy changes for phase transitions

You may be struck by one disturbing feature of the definition centred on Equation 7.2. It implies the transfer of heat to a substance without changing its temperature. In fact this is possible, but only under special circumstances.

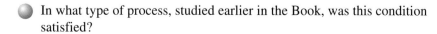 In what type of process, studied earlier in the Book, was this condition satisfied?

When heat is transferred to a substance undergoing an endothermic phase change, such as melting or boiling, the change takes place at constant temperature. One example (Section 8.2) is the vaporization of water at its boiling temperature, 100 °C (373.15 K at 1 atm pressure):

$$H_2O(l) = H_2O(g) \tag{8.4}$$

In this case, T is constant at 373.15 K, so we can calculate ΔS from Equation 7.2 if we can determine q, the heat that must be transferred to a sample of boiling water to convert it into steam. Our statement of the first law of thermodynamics enables us to do this:

$$\Delta H = q + w_{el} \tag{8.1}$$

If the conversion from water to steam were brought about by heat transfer (Figure 10.1, top), no electrical work would be done, so $\Delta H = q$. Alternatively, we could insulate the boiling water in, say, a Dewar flask and carry out the conversion by doing electrical work with an electrical heating element immersed in the water (Figure 10.1, bottom). This time, $q = 0$, and $\Delta H = w_{el}$, a relation that was used to determine ΔH in Question 8.2. Since enthalpy is a function of state, and the same change occurs in both cases, the two ΔH values are the same. This means that q, the heat supplied in the first conversion, is equal to w_{el}, the electrical work done in the second.

In this case therefore:

$$q = w_{el} \tag{10.1}$$

and, from Equation 7.2,

$$\Delta S = \frac{w_{el}}{T} \tag{10.2}$$

This equation holds *only* under the very special circumstances outlined above. In particular, the temperature T must not change, and no heat must be transferred into or out of the calorimeter. Under these conditions, all the electrical work done goes to bring about the desired phase change.

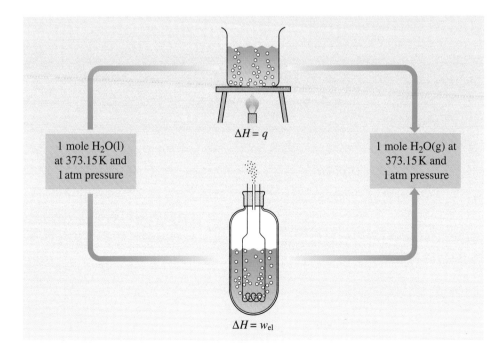

Figure 10.1
Two ways of converting 1 mole of water at the boiling point into steam. At the top, no electrical work is done so $\Delta H = q$; at the bottom, the sample is evaporated by electrical work in an insulated container when $q = 0$, so $\Delta H = w_{el}$. As the two values of ΔH are identical, $q = w_{el}$.

But the example in Question 8.2 highlights a further point. Evidently, *one* experiment suffices to determine *both* the enthalpy *and* entropy of vaporization for water (or any other pure substance), since, in the second of our two conversions (Figure 10.1, bottom), $\Delta H_{vap} = w_{el}$. Thus, ΔH_{vap} and ΔS_{vap} are related by a special case of Equation 10.2:

$$\Delta S_{vap} = \frac{\Delta H_{vap}}{T_{vap}} \qquad (10.3)$$

where T_{vap} is the normal boiling temperature. From the results in the answer to Question 8.2, the molar entropy of vaporization of water is then:

$$\Delta S_{vap} = \frac{44.3\,\text{kJ mol}^{-1}}{373.15\,\text{K}} = \frac{44.3 \times 10^3\,\text{J mol}^{-1}}{373.15\,\text{K}} = 118.7\,\text{J K}^{-1}\,\text{mol}^{-1}$$

Equation 10.3 is one example of a general relation between entropies and enthalpies of transition for pure substances.

Now try the following Question. This is important!

QUESTION 10.1

Solid chlorine can be made by cooling chlorine gas to, say, 150 K. On warming, the solid fuses (melts) at 172 K, when $\Delta H_{fus} = 6.4\,\text{kJ mol}^{-1}$. Further warming causes the liquid to boil (vaporize) at 239 K when $\Delta H_{vap} = 20.4\,\text{kJ mol}^{-1}$. One mole of chlorine ($Cl_2$) has a mass of 70.9 g. Calculate:

(a) the molar entropy of fusion of solid chlorine;

(b) the molar entropy of vaporization of liquid chlorine;

(c) the entropy of vaporization of 7.09 g of liquid chlorine.

10.3 Entropy changes of substances with temperature

In this Section, we shall seek a way of determining the entropy change of a substance when its temperature is increased. We shall use chlorine as an example, finding the entropy change when a sample of this substance is warmed from the absolute zero, 0 K, to 298.15 K (25 °C)

● Two important events happen during this temperature change: what are they?

● At 0 K, the chlorine is solid; it fuses at 172 K, and then vaporizes at 239 K (Question 10.1).

It follows that the entropy change of chlorine when it is warmed from 0 K to 298.15 K is the sum of five parts:

1   The entropy change of the solid chlorine when it is warmed from 0 K to the fusion temperature of 172 K.

2   The entropy of fusion of the solid chlorine at 172 K.

3   The entropy change of the liquid chlorine when it is warmed from 172 K to the vaporization temperature of 239 K.

4   The entropy of vaporization of the liquid chlorine at 239 K.

5   The entropy change of the gaseous chlorine when it is warmed from 239 K to 298.15 K.

In Question 10.1, you calculated parts 2 and 4; it remains to calculate parts 1, 3 and 5. In all three cases, the central problem is that of finding the entropy change when the temperature of the chlorine rises from a lower value, T_1, to a higher value, T_2. The first step is to determine q, the heat needed to bring about the temperature increase. The experimental procedure is similar to the one that we used when determining the entropy of a phase transition, and is shown in Figure 10.2. We insulate the chlorine in a calorimeter, and determine the electrical work, w_{el}, that is needed to bring about the required temperature change. The heat that will induce the same temperature increase when no electrical work is done can then be found from the equation

$$q = w_{el} \tag{10.1}$$

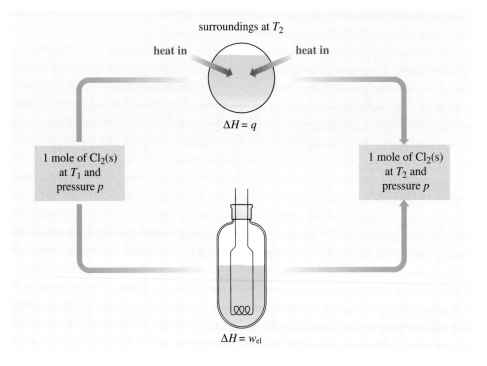

Figure 10.2
How to determine the heat q required to raise the temperature of one mole of solid chlorine from T_1 to T_2 at constant pressure. On the lower path, the job is done by a measured amount of electrical work in an insulated container in which $q = 0$, so $\Delta H = w_{el}$. On the top path, the job is done by heat transfer, so $w_{el} = 0$ and $\Delta H = q$. Thus, $q = w_{el}$.

Let us now consider an experiment in which a small temperature increase of 1.10 K is induced in a sample of solid chlorine at the very low temperature of 13.50 K. The results are shown in Table 10.1, and the value of w_{el} tells us that $q = 3.728$ J.

Table 10.1 Data for an experiment in which electrical work raises the temperature of one mole of solid chlorine from T_1 (= 13.50 K) to T_2 (= 14.60 K)

T_1/K	T_2/K	$\Delta T/K$	w_{el}/J	$T' = \dfrac{(T_1 + T_2)/2}{K}$	$\dfrac{\Delta S'}{J\ K^{-1}}$
13.50	14.60	1.10	3.728	14.05	0.265

But now we encounter the difficulty mentioned at the beginning of Section 10.1: Equation 7.2 shows that the entropy change is defined through a heat transfer at a constant temperature T:

$$\Delta S_{substance} = \frac{q}{T} \tag{7.2}$$

Yet the temperature has varied between 13.50 K and 14.60 K, so how can the equation be used?

If a single temperature must go into Equation 7.2, then the most sensible choice would seem to be the average of T_1 and T_2, which has been denoted T'. The small overall increase encourages this measure, because throughout the change, the temperature never differs by more than ±0.55 K from the mean value of 14.05 K. We can now use Equation 7.2 to calculate an approximate entropy change, $\Delta S'$, by using the equation

$$\Delta S' = \frac{q}{T'} = \frac{w_{el}}{T'} \tag{10.4}$$

As Table 10.1 records, this gives a value of $0.265\,\text{J K}^{-1}\,\text{mol}^{-1}$. Now, this suggests a way of estimating ΔS for a much bigger temperature change. Having raised the temperature to 14.60 K, we now induce a whole succession of further small temperature increases — from 14.60 K to 15.82 K, then from 15.82 K to 18.98 K, and so on. In Table 10.2, such a succession of increments takes the temperature from the original 13.50 K up to 90.00 K.

In all but one case, the approximate entropy change, $\Delta S'$, has been calculated from Equation 10.4. We could then add up all the $\Delta S'$ values to get an estimate of $\Sigma\Delta S'$, the total entropy change between 13.50 K and 90.00 K at 1 atm pressure:

$$S(Cl_2, \text{s}, 90.00\,\text{K}) - S(Cl_2, \text{s}, 13.50\,\text{K}) \approx \Sigma\Delta S' \tag{10.5}$$

Now try the following Question.

QUESTION 10.2

(a) To confirm that you have followed the argument, use the data in Table 10.2 to calculate the missing $\Delta S'$ value for the increment where T_1 is 20.64 K and T_2 is 25.22 K.

(b) Now use Equation 10.5 to obtain a value of the entropy difference between one mole of chlorine at 90.00 K and at 13.50 K. ADD UP COLUMN 6, AS'

ANSWER 39.393 JK⁻¹

For the solid chlorine of Question 10.2, the summation process yields a total entropy change of $39.39\,\text{J K}^{-1}\,\text{mol}^{-1}$. As Equation 10.5 implies, and as we have emphasized, this value is approximate, but there are mathematical techniques for improving it. To see how, look at the $\Delta S'$ values for the increments in column 6 of Table 10.2. They display no clear trend, sometimes rising, and sometimes falling as one moves down the column.

- Why does this happen?

- It is because the changes in temperature, ΔT, vary irregularly. When ΔT is large, $\Delta S'$ is large; when ΔT is small, $\Delta S'$ is small.

This suggests that we will observe a more regular pattern if we concentrate on $\Delta S'/\Delta T$. In the last column of Table 10.2, the values of $\Delta S'/\Delta T$ have been calculated for all but one of the increments.

- Fill in the missing value, using your answer to Question 10.2a. What pattern do you observe in the $\Delta S'/\Delta T$ values?

- The missing figure is $0.450\,\text{J K}^{-2}$. Values of $\Delta S'/\Delta T$ now increase steadily, reaching a maximum in the region of $T' = 40$ K; thereafter they decline.

Table 10.2 Data for an experiment in which successive amounts of electrical work raise the temperature of one mole of solid chlorine from 13.50 K to 90.00 K in a succession of small increments[*]

T_1/K	T_2/K	ΔT/K	w_{el}/J	$T' = \dfrac{(T_1 + T_2)/2}{K}$	$\dfrac{\Delta S'}{J\,K^{-1}}$	$\dfrac{\Delta S'/\Delta T}{J\,K^{-2}}$
13.50	14.60	1.10	3.728	14.05	0.265	0.241
14.60	15.82	1.22	4.676	15.21	0.307	0.252
15.82	18.98	3.16	17.598	17.40	1.011	0.320
18.98	20.64	1.66	12.794	19.81	0.646	0.389
20.64	25.22	4.58	47.275	22.93	2.062	0.450
25.22	27.52	2.30	30.716	26.37	1.165	0.506
27.52	32.24	4.72	77.927	29.88	2.608	0.553
32.24	35.64	3.40	68.340	33.94	2.014	0.592
35.64	39.72	4.08	92.538	37.68	2.456	0.602
39.72	45.02	5.30	133.449	42.37	3.150	0.594
45.02	49.26	4.24	118.415	47.14	2.512	0.592
49.26	54.26	5.00	149.805	51.76	2.894	0.579
54.26	62.92	8.66	285.486	58.59	4.873	0.563
62.92	67.92	5.00	175.350	65.42	2.680	0.536
67.92	73.08	5.16	188.257	70.50	2.670	0.517
73.08	77.42	4.34	162.937	75.25	2.165	0.499
77.42	82.00	4.58	176.316	79.71	2.212	0.483
82.00	86.12	4.12	162.505	84.06	1.933	0.469
86.12	90.00	3.88	155.461	88.06	1.765	0.455

[*] Note that a set of actual experimental results would show the electrical work done on the system as a whole: generally this would comprise a calorimeter and all its contents, including a thermometer and heating coil as well as the chlorine. The electrical work required to increase the temperature of the calorimeter, thermometer and heating coil is determined in a separate calibration experiment with an empty calorimeter. We have made allowance for everything except the chlorine in Table 10.2 by subtracting the electrical work done in this calibration experiment.

This trend is more clearly revealed by the plot of $\Delta S'/\Delta T$ against T' in Figure 10.3. Here, the important point is that because $\Delta S'/\Delta T$ shows a regular trend with temperature, a curve can be confidently drawn through the data points. Although the $\Delta S'$ values were obtained by an approximation, the curve-drawing process diminishes the effects of the errors in individual points. Moreover, the curve provides $\Delta S'/\Delta T$ at any value of T' between 13.50 K and 90.00 K, and this means that small temperature increments other than those in Table 10.2 can be studied.

We now take advantage of this by imagining an increment covering a very narrow temperature range. This increment has been drawn as a narrow vertical strip in the inset of Figure 10.4. The left-hand and right-hand verticals mark the initial and final temperatures, T_1 and T_2, respectively, and the width of the strip is ΔT. These features appear most clearly in the enlargements of the places where the strip intersects the

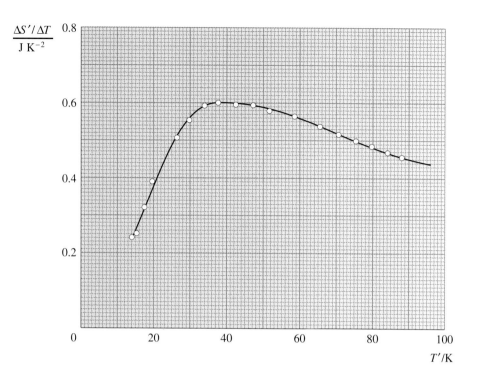

$\dfrac{\Delta S'/\Delta T}{\text{J K}^{-2}}$

Figure 10.3
A plot of the values of $\Delta S'/\Delta T$ against T' using the data on solid chlorine in Table 10.2.

curve and the T' axis. The mean temperature T' is marked by the central vertical, and intersects the curve at the value of $\Delta S'/\Delta T$ for the increment. To obtain the value of $\Delta S'$ for the increment, therefore, one simply multiplies this value of $\Delta S'/\Delta T$ by ΔT.

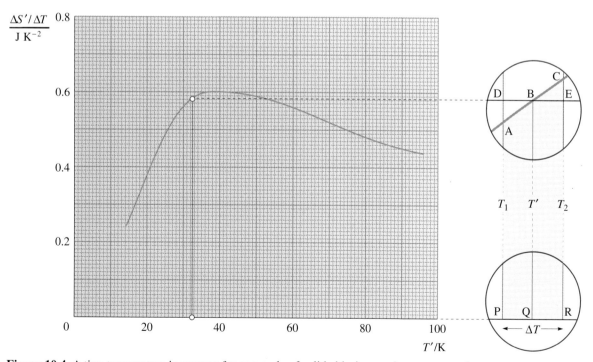

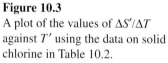

Figure 10.4 A tiny temperature increment for one mole of solid chlorine can be repre-sented by drawing a very narrow vertical strip in Figure 10.3, the vertical sides being at the lower and upper temperatures, T_1 and T_2. Here the intersections with the curve and the temperature axis have been magnified.

● So what area in Figure 10.4 is $\Delta S'$ equal to?

● The rectangle DPRE has an area DP × PR. As DP = BQ = $\Delta S'/\Delta T$, and ΔT = PR, this area is equal to $\Delta S'$.

Now this would be true whatever the width of the increment, but now for something that is not. If the increment is very narrow, the tiny section of curve AC that crosses it can be thought of as a straight line. Consequently, as B is the mid-point of both AC and DE, the area of the triangle BCE is equal to that of the triangle BAD.

● So what area besides the rectangle DPRE is also equal to $\Delta S'$?

● That of the strip ACRP beneath the curve. It can be obtained by adding BCE to the rectangle DPRE, and subtracting BAD. As the areas of the two triangles are the same, the rectangle and the strip have the same area.

Now suppose the sample of solid chlorine is taken through a much bigger increment — for example between the temperatures 20 K and 80 K in Table 10.2. These boundary temperatures can be represented by vertical lines from the horizontal axis to the curve in Figure 10.4. We can now imagine the temperature being increased from 20 K to 80 K by a whole succession of many tiny increments. The argument that we have just given shows that the entropy change for each tiny increment is the area of the strip beneath the curve bounded by the lower and upper temperatures. The entropy change for the overall large increment is the sum of the areas of these strips, and this is simply the area beneath the curve between the vertical lines at 20 K and 80 K.

In Figure 10.5, the curve of Figure 10.3 has been replotted with a dotted extension from 14.05 K, the lowest value of T' in Table 10.2, back to the absolute zero. Such an extension is always necessary in entropy measurement because the absolute zero is unattainable. It has been calculated by using a theoretical equation whose basis we do not discuss, but as you can see, it implies that $\Delta S'/\Delta T$ is zero at 0 K.

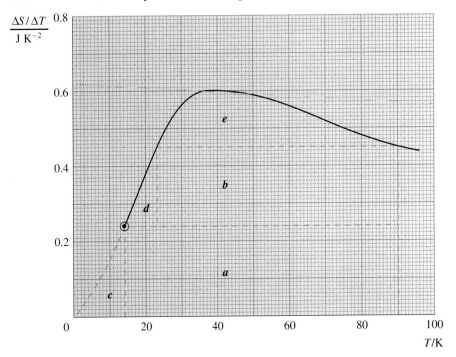

Figure 10.5
The plot of $\Delta S'/\Delta T$ against T for one mole of solid chlorine over the range 0–90 K. The extension of the experimental data from 14 K back to 0 K has been made using a theoretical equation. The area beneath the curve has been subdivided into approximate triangles and rectangles.

To check that you are following the argument, try the following Question.

QUESTION 10.3

Use Figure 10.5 to determine the molar entropy change for solid chlorine when its temperature rises from 0 K to 90 K. Assume that regions *a* and *b* in Figure 10.5 are rectangles, and that regions *c*, *d* and *e* are triangles whose area is

$$\frac{\text{base}}{2} \times \text{height}$$

To summarize, the entropy change for a pure substance over a temperature range that does not include a phase change can be determined from values of the quantity

$$\frac{\Delta S'}{\Delta T} = \frac{w_{el}}{T'\Delta T} \tag{10.6}$$

which is obtained for a succession of small increments of about 1–5 K, T' being the mean temperature of the increments, and w_{el} the electrical work needed to induce the temperature rise, ΔT. These values are plotted against T'; the entropy change is then the area beneath the curve between the lower and upper limits of the temperature range.

In Figure 10.5, we have dropped the superscript dashes against ΔS and T on the axes because we assume that the curve we derive will give us accurate values of $\Delta S/\Delta T$ for *very* small temperature increments centred on a temperature T.

Research workers who determine entropies perform the calculation mathematically, computing the area by a technique known as *integration*. Such workers write the entropy change when a solid substance increases its temperature from T_1 to T_2, in the form:

$$S(T_2) - S(T_1) = \int_{T_1}^{T_2} \frac{C_p}{T} \, dT \tag{10.7}$$

where C_p is the *heat capacity* of the substance, and the symbol $\int$ marks an integration process. Nevertheless, both this method, and the one we have used, amount to the same thing.

10.4 Absolute entropies: the third law of thermodynamics

By using measurements of the type listed in Table 10.2, and the technique used in answering Question 10.3, the entropy increment for *any* pure substance going from one temperature to another can be determined. But we can go a stage further. Unlike enthalpies, it is possible to determine an *absolute* value for the entropy of a pure substance at a given temperature. This absolute scale is based on a third, and final, 'rule of the game' — a proposition known as the **third law of thermodynamics**. This law is adopted because it establishes useful relationships between important ideas in thermodynamics and in quantum mechanics. For our purposes, we shall interpret the law as saying:

The entropy of a pure crystalline substance is zero in the limit as the temperature approaches absolute zero $(T \to 0)$ *.

* The absolute zero of temperature cannot actually be obtained, but it can be approached (experimentally): this is why we write $T \to 0$.

79

If a pure solid substance undergoes no phase change between 0 and 298.15 K, its absolute entropy at 298.15 K can now be determined from a plot like that in Figure 10.5. We know that the area under the graph of $\Delta S'/\Delta T$ against T between 0 and 298.15 K is given by

$$\text{area} = S(298.15\,\text{K}) - S(T \rightarrow 0) \tag{10.8}$$

But the third law tells us that $S(T \rightarrow 0)$ is zero, so Equation 10.8 becomes

$$\text{area} = S(298.15\,\text{K}) \tag{10.9}$$

In other words, the area under the graph represents the *absolute* entropy of the substance at 298.15 K. You should be able to see that the value you obtained in answering Question 10.3 represents the absolute molar entropy of solid chlorine at 90 K.

As with enthalpies, calorimetrically determined entropies must be adjusted to a standard reference state, and this requires the specification of a standard value, $p^{\ominus}$, of pressure: once again the value chosen is 100 kPa (or one bar), which is a typical atmospheric pressure. Strictly speaking, the symbol for a standard absolute molar entropy should be written $S_{m}^{\ominus}$ together with a specification of the temperature. However, the subscript m is generally omitted where doing so does not cause any confusion. Indeed, it is normal to refer simply to 'absolute entropies'.

10.5 Absolute entropy of chlorine gas at 298.15 K

At the beginning of Section 10.2 we set ourselves the task of determining the entropy change ΔS of a sample of chlorine when it is warmed from 0 K to 298.15 K. The third law of thermodynamics now tells us that this value of ΔS is identical with the absolute entropy, $S^{\ominus}$, of chlorine at 298.15 K. If chlorine underwent no phase change between 0 K and 298.15 K, the absolute entropy could be obtained by measuring the area under the graph of $\Delta S'/\Delta T$ against T; in Question 10.3, $S^{\ominus}(Cl_2, s)$ at 90 K was obtained in just this way. However, chlorine melts at 172 K and vaporizes at 239 K. Let us see how this affects the entropy determination by examining the complete plot of $\Delta S/\Delta T$ against T for chlorine between 0 and 298.15 K, which is shown in Figure 10.6.

⬤ What is the significance of the breaks at 172 K and 239 K?

⬤ The one at 172 K marks the fusion temperature, where solid changes to liquid; the one at 239 K marks the vaporization temperature, where liquid changes to gas.

The fusion and vaporization breaks divide the area beneath the curve into three parts labelled 1, 3 and 5. Area 1 represents the entropy change when the mole of solid chlorine is warmed from 0 K to 172 K; area 3, the entropy change when one mole of liquid chlorine is warmed from 172 K to 239 K, and area 5 the entropy change when one mole of gaseous chlorine is warmed from 239 K to 298.15 K.

⬤ Is the sum of these three areas equal to the value of the absolute entropy of chlorine gas, $S^{\ominus}(Cl_2, g)$ at 298.15 K?

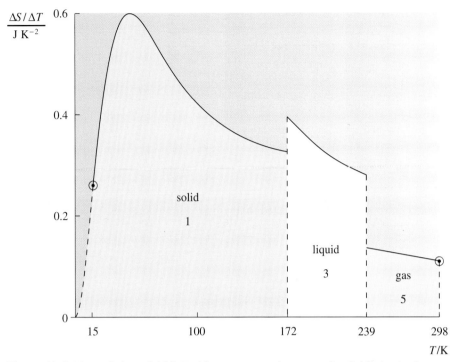

Figure 10.6 The variation of $\Delta S/\Delta T$ with temperature for one mole of chlorine in the range from $T \to 0$ to $T = 298.15$ K. The points mark the limits of the temperature range over which *experimental* measurements are made.

⬤ No; look back to the beginning of Section 10.3, where we divided the entropy change when one mole of chlorine is warmed from 0 to 298.15 K into five stages; areas 1, 3 and 5 only give us the entropy changes for stages 1, 3 and 5.

The calculation of $S^{\ominus}(Cl_2, g)$ at 298.15 K can therefore be completed by adding the entropy changes for steps 2 and 4. These are the ΔS values for the fusion and vaporization of the chlorine, which were calculated in Question 10.1. The addition for the five stages is shown in Table 10.3; the final value of $S^{\ominus}(Cl_2, g)$ at 298.15 K is 223.1 J K^{-1} mol^{-1}, as given in the *Data Book* (on the CD-ROM associated with this Book).

Table 10.3 The absolute entropy of one mole of Cl_2 at 298.15 K and 100 kPa is obtained as a sum of the entropy changes of five stages as the chlorine sample is warmed from $T \to 0$ to $T = 298.15$ K

Stage		ΔS/J K^{-1} mol^{-1}
1	$Cl_2(s)$ warmed from $T \to 0$ to 172 K	70.7
2	$Cl_2(s)$ fusion at 172 K	37.2
3	$Cl_2(l)$ warmed from 172 K to 239 K	21.9
4	$Cl_2(l)$ vaporization at 239 K	85.4
5	$Cl_2(g)$ warmed from 239 K to 298.15 K	7.9
6	$S^{\ominus}(Cl_2, g)$ at 298.15 K	223.1

10.6 Absolute entropies in the *Data Book*

In general then, the absolute entropy of a pure substance at a temperature T may be obtained by measuring the area beneath the graph of $\Delta S/\Delta T$ against T between 0 K and T, and adding the terms $\Delta H_{fus}/T_{fus}$ and $\Delta H_{vap}/T_{vap}$, where these are relevant. Notice that this extension carries with it an important implication: the absolute entropy of a gas will contain contributions from the entropies of fusion and vaporization. Not surprisingly, therefore, the entropies of gases are larger than those of comparable liquids or solids. We shall return to this point in the next Section.

In summary: once the necessary measurements and calculations have been performed, the resulting values of $S^{\ominus}$ are listed in the chemical literature. A compilation of such values at 298.15 K is included in the *Data Book*. A quick glance through these values should confirm the generalization we made above. For instance, the values in Table 10.4 confirm the tendency for the entropies of gases to be larger than those of solids or liquids.

Among the entries in the *Data Book*, you will also find values for individual aqueous ions. Like enthalpy data, these rest on the arbitrary assignment of a zero value to one ion: again, the hydrogen ion is chosen, so (by definition)

$$S^{\ominus}(\text{H}^+, \text{aq}, 298.15\ \text{K}) = 0 \qquad (10.10)$$

This is another example of an *arbitrary* convention [*]; once again, the values so obtained are internally consistent.

Table 10.4
The absolute entropies of gases at 298.15 K are usually larger than those of solids or liquids, as these values show

Substance	$S^{\ominus}/\text{J K}^{-1}\,\text{mol}^{-1}$
Ag(s)	42.6
Ag(g)	173.0
I_2(s)	116.1
I_2(g)	260.7
Br_2(l)	152.2
Br_2(g)	245.5

10.7 The entropy change for a reaction

The values of $S^{\ominus}$ in the *Data Book* can be used to calculate the standard molar entropy change for a reaction, $\Delta S_m^{\ominus}$, by substituting them into a more precise version of Equation 7.4 (Section 7.1):

$$\Delta S_m^{\ominus} = \Sigma S^{\ominus}(\text{products}) - \Sigma S^{\ominus}(\text{reactants}) \qquad (10.11)$$

As before, the symbol $\Sigma S^{\ominus}$(products) means the sum of the absolute entropies of the products, etc. As with the corresponding expression for enthalpies in Equation 9.18 (Section 9.3.2), the individual contributions to these sums must first be multiplied by the numbers that precede them in the balanced reaction equation.

QUESTION 10.4

Using information from the *Data Book*, calculate $\Delta S_m^{\ominus}$ for each of the following reactions at 298.15 K:

(a) (i) $\text{Na(s)} + \frac{1}{2}\text{Cl}_2\text{(g)} = \text{NaCl(s)}$ (5.2)

   (ii) $\text{Mg(s)} + \frac{1}{2}\text{O}_2\text{(g)} = \text{MgO(s)}$ (2.5)

   (iii) $\text{CaCO}_3\text{(s)} = \text{CaO(s)} + \text{CO}_2\text{(g)}$ (9.10)

   (iv) $\text{N}_2\text{O}_4\text{(g)} = 2\text{NO}_2\text{(g)}$ (9.19)

(b) (v) $\text{Zn(s)} + 2\text{H}^+\text{(aq)} = \text{Zn}^{2+}\text{(aq)} + \text{H}_2\text{(g)}$ (2.4)

   (vi) $\text{Cu(s)} + 2\text{H}^+\text{(aq)} = \text{Cu}^{2+}\text{(aq)} + \text{H}_2\text{(g)}$ (2.9)

The examples in Question 10.4 underline important differences between entropy and enthalpy data — points that will be taken up in the next Section. But closer inspection of

[*] It is only because of this arbitrary convention that the absolute entropies of some aqueous ions have negative values; for example, $S^{\ominus}(\text{Mg}^{2+}, \text{aq}) = -138\ \text{J K}^{-1}\,\text{mol}^{-1}$.

the results from Question 10.4a raises a further point. In each case the *sign* of $\Delta S_m^{\ominus}$ for the reaction can be correlated with whether it results in an increase or a decrease in the number of moles of *gaseous* species. This observation, which is supported by the more extensive compilation in Table 10.5, harks back to our earlier comment about the large $S^{\ominus}$ values for gases. It suggests an important and useful generalization:

> If a reaction involving only pure substances results in an increase in the number of moles of gaseous species, then $\Delta S_m^{\ominus}$ will be positive; if the result is a decrease in the number of moles of gaseous species, then $\Delta S_m^{\ominus}$ will be negative.

Table 10.5 Values of $\Delta S_m^{\ominus}$ at 298.15 K for a variety of reactions

Reaction	$\Delta S_m^{\ominus}$ /J K^{-1} mol^{-1}
$I_2(s) = I_2(g)$	144.6
$N_2(g) + 3H_2(g) = 2NH_3(g)$	−198.7
$H_2O(l) = H_2O(g)$	118.9
$Na(g) + Cl(g) = NaCl(s)*$	−246.8
$PbO(s) + C(s) = Pb(s) + CO(g)$	190.3

* Note that chlorine *atoms* are involved in this reaction.

But note the qualification in this generalization: it is restricted to changes in which every reactant and product is a pure substance. In the reactions in Question 10.4b, hydrogen gas is produced, but $\Delta S_m^{\ominus}$ is negative! Reactions in solution, especially those involving aqueous ions, are distinctly more complex, and there is then no simple way of predicting the sign of $\Delta S_m^{\ominus}$ with any certainty.

10.8 Summary of Section 10

1 The entropy of transition of a pure substance is $\Delta H/T$, where ΔH is the enthalpy of transition at the transition temperature T.

2 The entropy change of a pure substance when its temperature is raised from T_1 to T_2 can be determined from values of the quantity $w_{el}/T'\Delta T$ obtained for a succession of *small* increments, ΔT, of about 1–5 K, T' being the mean temperature of the increment, and w_{el} the electrical work needed to induce the increment ΔT. These values are plotted against T'; the required entropy change is the area beneath the curve between T_1 and T_2, plus values of $\Delta H/T$ for any phase transitions in this range.

3 The procedure of point 2 allows absolute entropies to be determined, because according to the third law of thermodynamics, when the temperature of a substance is raised from 0 K to T, the entropy change is equal to the absolute entropy of the substance at temperature T.

4 The entropy change of a reaction, $\Delta S_m^{\ominus}$, can be obtained from the absolute entropies, $S^{\ominus}$, by using the equation

$$\Delta S_m^{\ominus} = \Sigma S^{\ominus}(\text{products}) - \Sigma S^{\ominus}(\text{reactants}) \qquad (10.11)$$

5 Because the entropies of gases are so large, a reaction between pure substances that results in an increase in the number of moles of gaseous species will nearly always have a positive value of $\Delta S_m^{\ominus}$.

THE GIBBS FUNCTION 11

In Section 7, we used the second law of thermodynamics to show that, for a spontaneous change at constant temperature and pressure,

$$\Delta H - T\Delta S < 0 \tag{7.10}$$

In Sections 8 and 9, you were shown how ΔH could be determined, and in Section 10 how the same could be done for ΔS. The values of ΔH and ΔS then allow you to predict whether or not a given reaction can occur, without ever trying it in practice. Such is the power of thermodynamics!

The two-term quantity, $\Delta H - T\Delta S$, is so important that it deserves a symbol of its own. We define a quantity G in terms of enthalpy and entropy, writing it

$$G = H - TS \tag{11.1}$$

G is known variously as the **Gibbs function** or the Gibbs free energy, after the American physicist, Josiah Willard Gibbs (Figure 11.1).

○ According to this definition, what is the SI unit of G?

○ The SI unit of H is J, and that of TS is $K \times J\,K^{-1} = J$, so the SI unit of G is the joule.

Thus, a change in the Gibbs function at a constant temperature T (and implicitly at constant pressure) is given by

$$\Delta G = \Delta H - T\Delta S \tag{11.2}$$

Comparing this result with the inequality 7.10, leads to the following simplified criterion for a spontaneous reaction:

$$\Delta G < 0 \text{ (at constant } T \text{ and } p) \tag{11.3}$$

○ Does this inequality accord with the intuition about an 'uphill–downhill' character to chemical reactions, as developed in Section 5?

○ Yes. In terms of our simple analogy, it seems that spontaneous reactions really do 'roll downhill', but on a landscape where 'altitude' is measured by the Gibbs function, not the enthalpy (as in Thomsen's hypothesis).

Notice that when reactants and products are in their standard states, ΔG is written $\Delta G^{\ominus}$, and the **standard molar Gibbs function change** is written $\Delta G_{m}^{\ominus}$. Thus, the values of $\Delta H_{m}^{\ominus}$ and $\Delta S_{m}^{\ominus}$ calculated from tabulated data are strictly related to $\Delta G_{m}^{\ominus}$, as defined by an expression analogous to Equation 11.2:

$$\Delta G_{m}^{\ominus} = \Delta H_{m}^{\ominus} - T\Delta S_{m}^{\ominus} \tag{11.4}$$

> We shall take the criterion $\Delta G_{m}^{\ominus} < 0$ as our definition of a thermodynamically favourable reaction; conversely, a reaction is considered to be unfavourable if $\Delta G_{m}^{\ominus} > 0$.

Figure 11.1
Josiah Willard Gibbs (1839–1903). Apart from three years study in Europe during his twenties, this taciturn bachelor lived his life in a house close to Yale University, where he was first a student and then an unsalaried professor. In 1863, he received the first US engineering doctorate for a thesis on the teeth of wheels in spur gearing. During the 1870s, he applied the second law of thermodynamics to chemical equilibrium in a series of papers which later revolutionized physical chemistry. Johns Hopkins University at Baltimore then offered him a job, and Yale University at last produced a salary to retain his allegiance.

Now try the following Question.

QUESTION 11.1

(a) Use information from the *Data Book* to calculate the standard molar entropy change of the reaction

$$SOCl_2(l) + H_2O(l) = SO_2(g) + 2HCl(g) \qquad (6.4)$$

(b) The value of $\Delta H_m^{\ominus}$ for this process was given in Section 6 as $47.1\,\text{kJ mol}^{-1}$. Calculate the value of $\Delta G_m^{\ominus}$.

(c) Are the values of $\Delta H_m^{\ominus}$, $\Delta S_m^{\ominus}$ and $\Delta G_m^{\ominus}$ in accord with the observation that Reaction 6.4 occurs readily at room temperature and pressure?

11.1 The Gibbs function and the equilibrium constant

To adopt the criterion $\Delta G_m^{\ominus} < 0$ as the mark of a thermodynamically favourable reaction is to select a special case of inequality 11.3. The great advantage of concentrating on $\Delta G_m^{\ominus}$ is that this quantity is directly related to the equilibrium constant, K, the other measure of how favourable a reaction is.

● When a reaction is very favourable, is the equilibrium constant K large or small?

● It is large; for example, for a general reaction such as

$$A + B = C + D \qquad (11.5)$$

$$K = \frac{[C][D]}{[A][B]} \qquad (11.6)$$

If the reaction is favourable, the products C and D must predominate at equilibrium, so K must be large.

From the discussion above, $\Delta G_m^{\ominus}$ must be negative for a reaction like this: conversely, if $\Delta G_m^{\ominus}$ is positive, then K must be small. These observations suggest strongly that $\Delta G_m^{\ominus}$ and K must be related in some way. Unfortunately, the derivation of this important relation lies beyond the scope of this Book, so we must ask you to take on trust that it is as follows:

$$\Delta G_m^{\ominus} = -RT \ln K \qquad (11.7)^*$$
$$= -2.303RT \log K \qquad (11.8)^*$$

where R is a constant, known as the **gas constant**, which has the value $R = 8.314\,\text{J K}^{-1}\text{mol}^{-1}$. The term $\ln K$ is the natural logarithm of K; $\log K$ is its logarithm to the base 10.

Although we cannot prove the expressions in Equations 11.7 and 11.8 here, they certainly fulfil the conditions outlined above. Consider, for example, the values of $\log K$ and of K for a range of values of $\Delta G_m^{\ominus}$ given in Table 11.1. In particular, notice that when $\Delta G_m^{\ominus}$ is negative, K is greater than one and, when $\Delta G_m^{\ominus}$ is positive, K is less than one.

* See Maths Help Box overleaf.

Table 11.1 Some numerical values of $\Delta G_m^{\ominus}$ with corresponding values of log K and K at 298.15 K (from Equation 11.8, with $R = 8.314\,\mathrm{J\,K^{-1}\,mol^{-1}}$)

$\Delta G_m^{\ominus}/\mathrm{kJ\,mol^{-1}}$	log K	K
−500	87.6	3.8×10^{87}
−100	17.5	3.3×10^{17}
−50	8.76	5.7×10^8
−10	1.75	5.6×10
−5	0.876	7.5
0	0	1
5	−0.876	1.3×10^{-1}
10	−1.75	1.8×10^{-2}
50	−8.76	1.7×10^{-9}
100	−17.5	3.0×10^{-18}
500	−87.6	2.6×10^{-88}

MATHS HELP
Logarithms of equilibrium constants

Equation 11.8 differs from Equation 11.7 only in the additional factor 2.303, and in the subtitution of logarithms to the base ten (log) for natural logarithms (ln). These differences arise because, for any number x, the two kinds of logarithm are related as $\ln x = 2.303 \log x$. Scientific calculators include buttons for both ln and log, so should you need to calculate $\Delta G_m^{\ominus}$ from K, you can use either Equation 11.7 or Equation 11.8. Note that the equilibrium constants calculated from $\Delta G_m^{\ominus}$ values in Table 11.1 are dimensionless, whereas you are used to equilibrium constants with dimensions of concentration (e.g. Question 4.2). For a comment on this, see the Appendix.

The values of log K and K reveal a key feature of logarithms to the base 10. When a number (in this case K) is written down in scientific notation, log K is close to the power to which 10 has to be raised.

QUESTION 11.2

Consider a simple isomerization reaction of the type A = B. If $\Delta G_m^{\ominus} = 0$, what proportion of the reactant, A, remains at equilibrium?

It seems, then, that $\Delta G_m^{\ominus}$ is really just a measure of the equilibrium constant for a reaction at a particular temperature. Table 11.1 shows that the more negative $\Delta G_m^{\ominus}$ is, the larger is K and so the more favourable is the reaction. Indeed, the value of $\Delta G_m^{\ominus}$ at 298.15 K need be only moderately negative (around −50 kJ mol⁻¹) to signify a reaction for which the equilibrium position strongly favours the products. By contrast, an equally positive value of $\Delta G_m^{\ominus}$ indicates a reaction for which little reactant has been converted by the time equilibrium is attained.

To reinforce this idea, try the following Question.

QUESTION 11.3

Use Equation 11.4 to calculate $\Delta G_m^{\ominus}$ at 298.15 K for each of the following reactions:

(i) $Na(s) + \frac{1}{2}Cl_2(g) = NaCl(s)$ (5.2)

(ii) $Cu(s) + 2H^+(aq) = Cu^{2+}(aq) + H_2(g)$ (2.9)

(ii) $Zn(s) + 2H^+(aq) = Zn^{2+}(aq) + H_2(g)$ (2.4)

Decide whether or not the reactions are thermodynamically favourable under standard conditions at 298.15 K.

11.2 Final survey of the thermodynamic database

We are now almost ready for the promised thermodynamic analysis of the hypothesis raised in Sections 4 – 4.2. However, before we embark on it, you must be able to exploit the thermodynamic database in the *Data Book*; this Section should ensure that this is the case. The database consists of a table, which lists three thermodynamic quantities, $\Delta H_f^{\ominus}$, $\Delta G_f^{\ominus}$ and $S^{\ominus}$ for a wide range of chemical substances. A small selection is given in Table 11.2. Two kinds of material are represented: substances that have no overall charge, and aqueous ions.

Let us consider substances with no overall charge first. Sodium chloride, NaCl(s), is a good example.

● Write the equation to which $\Delta H_f^{\ominus}$ for NaCl(s) refers.

● Section 9.3.1 identified $\Delta H_f^{\ominus}$ for a pure substance as the value of $\Delta H_m^{\ominus}$ for the reaction in which the substance is formed from its elements in their standard reference states:

$Na(s) + \frac{1}{2}Cl_2(g) = NaCl(s)$ (5.2)

Table 11.2 Examples of thermodynamic data at 298.15 K from the *Data Book*

Substance	State	$\dfrac{\Delta H_f^{\ominus}}{kJ\,mol^{-1}}$	$\dfrac{\Delta G_f^{\ominus}}{kJ\,mol^{-1}}$	$\dfrac{S^{\ominus}}{J\,K^{-1}\,mol^{-1}}$
Na	s	0	0	51.2
Na$^+$	aq	−240.1	−261.9	59.0
NaCl	s	−411.2	−384.2	72.1
Cl$_2$	g	0	0	223.1
Cl$^-$	aq	−167.2	−131.2	56.5
H$_2$	g	0	0	130.7
H$^+$	aq	0	0	0
Zn	s	0	0	41.6
Zn^{2+}	aq	−153.9	−147.0	−112.1

You can see that $\Delta G_f^{\ominus}$ also carries a subscript f, and this marks the fact that it too refers to this type of reaction.

For an uncharged substance, $\Delta G_f^{\ominus}$, the standard Gibbs function of formation, is the value of $\Delta G_m^{\ominus}$ for the reaction in which the substance is formed from its elements as in Equation 5.2. As with $\Delta H_f^{\ominus}$ values, this means that $\Delta G_f^{\ominus}$ for any element in its standard reference state is zero. Values of $\Delta G_f^{\ominus}$ can now be obtained by using the equation

$$\Delta G_f^{\ominus} = \Delta H_f^{\ominus} - T\Delta S_f^{\ominus} \qquad (11.9)$$

which is just a special case of Equation 11.4.

(5.2) $Na_{(s)} + \frac{1}{2} Cl_{2(g)} = NaCl_{(s)}$

This expression, however, contains a trap for the unwary, so *be careful*. The entropy change in Equation 11.9 is $\Delta S_f^{\ominus}$, but the tables contain absolute entropies $S^{\ominus}$. To use Equation 11.9, you must calculate $\Delta S_f^{\ominus}$ for the substance from the values of $S^{\ominus}$ in the data base.

● Use the $S^{\ominus}$ values in Table 11.2 to calculate $\Delta S_f^{\ominus}$ for NaCl(s).

● $\Delta S_f^{\ominus}$ refers to the formation reaction:

$$Na(s) + \frac{1}{2}Cl_2(g) = NaCl(s) \qquad (5.2)$$

So using Equation 10.11:

$$\Delta S_f^{\ominus} = S^{\ominus}(NaCl, s) - S^{\ominus}(Na, s) - \frac{1}{2}S^{\ominus}(Cl_2, g)$$
$$= (72.1 - 51.2 - \frac{1}{2} \times 223.1)\,J\,K^{-1}\,mol^{-1}$$
$$= -90.7\,J\,K^{-1}\,mol^{-1}$$

As $\Delta H_f^{\ominus} = -411.2\,kJ\,mol^{-1}$, Equation 11.9 tells us that:
$$\Delta G_f^{\ominus} = -411.2\,kJ\,mol^{-1} - (298.15\,K) \times (-90.7\,J\,K^{-1}\,mol^{-1})$$

Remember to allow for the fact that *joules* appear in the units of ΔS and *kilojoules* in the units of ΔH:
$$\Delta G_f^{\ominus} = -411.2\,kJ\,mol^{-1} + 27\,040\,J\,mol^{-1}$$
$$= -411.2\,kJ\,mol^{-1} + 27.0\,kJ\,mol^{-1}$$
$$= -384.2\,kJ\,mol^{-1}$$

This is the value given in the Tables, and the calculation establishes the nature of the relationship between $\Delta H_f^{\ominus}$, $\Delta G_f^{\ominus}$ and $S^{\ominus}$ for a pure substance.

In the case of aqueous ions, we have not yet defined the formation reaction explicitly. Instead, in Section 9.3.3, we obtained $\Delta H_f^{\ominus}$ values for individual ions by assigning an arbitrary value to one ion. This assignment was made by defining

$$\Delta H_f^{\ominus}(H^+, aq) = 0 \qquad (9.24)$$

However, it is not difficult to show that this assignment is equivalent to defining $\Delta H_f^{\ominus}$ for a positive aqueous ion as $\Delta H_m^{\ominus}$ for the reaction in which the ion is formed by the reaction of its elements in their standard reference states with $H^+(aq)$, the other product being hydrogen gas. Thus, for zinc, this reaction is

$$Zn(s) + 2H^+(aq) = Zn^{2+}(aq) + H_2(g) \qquad (2.4)$$

Equation 9.18 then tells us that

$$\Delta H_m^{\ominus} = \Delta H_f^{\ominus}(Zn^{2+}, aq) + \Delta H_f^{\ominus}(H_2, g) - \Delta H_f^{\ominus}(Zn, s) - 2\Delta H_f^{\ominus}(H^+, aq)$$

But by definition, values of $\Delta H_f^{\ominus}$, for $H_2(g)$, $Zn(s)$ and $H^+(aq)$ are zero, so

$$\Delta H_m^{\ominus} = \Delta H_f^{\ominus}(Zn^{2+}, aq)$$

For negative aqueous ions, the formation reaction is similarly defined, but now $H^+(aq)$ goes on the right and hydrogen gas on the left; for example

$$\tfrac{1}{2}Cl_2(g) + \tfrac{1}{2}H_2(g) = Cl^-(aq) + H^+(aq) \tag{9.22}$$

Again, all values of $\Delta H_f^{\ominus}$ are zero by definition, except the one for $Cl^-(aq)$.

⬤ Write the formation reaction for $S^{2-}(aq)$

⬤ The equation is

$$S(s) + H_2(g) = S^{2-}(aq) + 2H^+(aq) \tag{11.10}$$

Like values of $\Delta H_f^{\ominus}$, values of $S^{\ominus}$ (as you saw in Section 10.5) and of $\Delta G_f^{\ominus}$ require arbitrary assignments to one ion before a set of individual values can be recorded. Again, zero values are assigned to the hydrogen ion. Hence we arbitrarily say that:

$$S^{\ominus}(H^+, aq, 298.15\,K) = 0 \tag{10.10}$$
$$\Delta G_f^{\ominus}(H^+, aq, 298.15\,K) = 0 \tag{11.11}$$

This ensures that, for aqueous ions as for other substances, $\Delta G_f^{\ominus}$ and $\Delta H_f^{\ominus}$ refer to the same formation reaction, and that they are, respectively, the values of $\Delta G_m^{\ominus}$ and $\Delta H_m^{\ominus}$ for that formation reaction.

As for pure substances, when performing calculations with Equation 11.9, you must remember that the equation includes $\Delta S_f^{\ominus}$, and not $S^{\ominus}$.

⬤ Use the $S^{\ominus}$ values in Table 11.2 to calculate $\Delta S_f^{\ominus}$ for $Na^+(aq)$.

⬤ The formation reaction is:

$$Na(s) + H^+(aq) = Na^+(aq) + \tfrac{1}{2}H_2(g) \tag{9.26}$$
$$\Delta S_f^{\ominus} = S^{\ominus}(Na^+, aq) + \tfrac{1}{2}S^{\ominus}(H_2, g) - S^{\ominus}(Na, s) - S^{\ominus}(H^+, aq)$$
$$= (59.0 + \tfrac{1}{2} \times 130.7 - 51.2 - 0)\,J\,K^{-1}\,mol^{-1}$$
$$= 73.2\,J\,K^{-1}\,mol^{-1}$$

As $\Delta H_f^{\ominus} = -240.1\,kJ\,mol^{-1}$, Equation 11.9 tells us that:

$$\Delta G_f^{\ominus} = -240.1\,kJ\,mol^{-1} - (298.15\,K) \times 73.2\,J\,K^{-1}\,mol^{-1}$$
$$= -261.9\,kJ\,mol^{-1}$$

Again, this is the value given in the Tables.

Finally, for any balanced reaction equation, the value of $\Delta H_m^{\ominus}$, $\Delta S_m^{\ominus}$ or $\Delta G_m^{\ominus}$ at 298.15 K can be obtained by doing the simple sums implied by the following equations (always remembering that the individual contributions to these sums must first be multiplied by the numbers preceding them in the equation):

$$\Delta H_m^{\ominus} = \Sigma \Delta H_f^{\ominus}(products) - \Sigma \Delta H_f^{\ominus}(reactants) \tag{9.18}$$
$$\Delta S_m^{\ominus} = \Sigma S^{\ominus}(products) - \Sigma S^{\ominus}(reactants) \tag{10.11}$$
$$\Delta G_m^{\ominus} = \Sigma \Delta G_f^{\ominus}(products) - \Sigma \Delta G_f^{\ominus}(reactants) \tag{11.12}$$

QUESTION 11.4

In Question 11.1, you calculated $\Delta G_m^{\ominus}$ for the reaction

$$SOCl_2(l) + H_2O(l) = SO_2(g) + 2HCl(g) \tag{6.4}$$

The value was obtained by computing the values of $\Delta H_m^{\ominus}$ and $\Delta S_m^{\ominus}$ using Equations 9.18 and 10.11, and then substituting them into Equation 11.4. Now repeat the calculation using values of $\Delta G_f^{\ominus}$ from the *Data Book* and Equation 11.12. Check that you get the same answer.

11.3 Summary of Section 11

1   The standard Gibbs function change for a chemical reaction at a constant temperature T is given by

$$\Delta G_m^{\ominus} = \Delta H_m^{\ominus} - T\Delta S_m^{\ominus} \tag{11.4}$$

2   At constant temperature and pressure, the criterion for a spontaneous change is $\Delta G < 0$. We take the criterion $\Delta G_m^{\ominus} < 0$ as a definition of a thermodynamically favourable reaction; conversely, a reaction is considered to be unfavourable if $\Delta G_m^{\ominus} > 0$.

3   At constant temperature, the more negative is $\Delta G_m^{\ominus}$, the larger is the equilibrium constant, and the more favourable is the reaction.

4   The *Data Book* (on the CD-ROM) contains a table of $\Delta H_f^{\ominus}$, $\Delta G_f^{\ominus}$ and $S^{\ominus}$ values for chemical substances. For pure substances, $\Delta H_f^{\ominus}$ and $\Delta G_f^{\ominus}$ refer to a formation reaction in which the substance is formed from its elements in their standard reference states. For aqueous ions, the formation reaction takes one of two forms: formation of cations from their elements and $H^+(aq)$ with production of hydrogen, or formation of anions from their elements and $H_2(g)$ with production of $H^+(aq)$.

5    $\Delta H_f^{\ominus}$ and $\Delta G_f^{\ominus}$ are related by a special case of Equation 11.4:

$$\Delta G_f^{\ominus} = \Delta H_f^{\ominus} - T\Delta S_f^{\ominus} \tag{11.9}$$

To use this equation, $\Delta S_f^{\ominus}$ values have to be calculated from the $S^{\ominus}$ values in tables.

6    $\Delta G_m^{\ominus}$ values for balanced chemical equations can be obtained from

$$\Delta G_m^{\ominus} = \Sigma \Delta G_f^{\ominus}(products) - \Sigma \Delta G_f^{\ominus}(reactants) \tag{11.12}$$

QUESTION 11.5

This question is concerned with the information in Table 11.3. Taking any further information you require from the *Data* Book, fill in the two blank entries in Table 11.3. Is the formation of $Sc_2O_3(s)$ from the elements thermodynamically favourable under standard conditions at 298.15 K?

Table 11.3 Thermodynamic data at 298.15 K for scandium (Sc)

Substance	State	$\dfrac{\Delta H_f^{\ominus}}{kJ\ mol^{-1}}$	$\dfrac{\Delta G_f^{\ominus}}{kJ\ mol^{-1}}$	$\dfrac{S^{\ominus}}{J\ K^{-1}\ mol^{-1}}$
Sc	s	0	0	34.6
Sc^{3+}	aq		−586.6	−255.2
Sc_2O_3	s	−1 908.8		77.0

METALS AND THEIR EASE OF OXIDATION

12

Now that our introduction to some of the basic ideas of chemical thermodynamics is complete, we shall remind you of some of the problems and questions that we elicited earlier in Sections 2–4 of this Book.

Sections 2.2–2.4 discussed and observed the reactions of metals with $H^+(aq)$, with the aqueous ions of other metals, with oxygen and with the halogens; Section 3 briefly discussed the extraction of metals from their ores. In Section 4, we elicited a plausible hypothesis from these experiments and observations. It seemed as if it might be possible to arrange metals in order of their reactivity or their tendency to be oxidized. This order looked as though it might possibly remain the same even when we changed the oxidizing agent.

In Section 4.1 we looked at this hypothesis more critically. Firstly, we noted that none of the experiments or observations on which the hypothesis was based was quantitative. Secondly, we scrutinized the language in which the hypothesis was expressed, in particular the concepts of *reactivity* and *tendency to be oxidized*; neither concept took account of the crucial distinction between the rate of a reaction and its equilibrium position.

In Section 4.1, we remarked that both these criticisms could be alleviated by comparing equilibrium constants for the reactions of metals with a particular oxidizing agent. We have now provided you with the means to do this. In Section 11.1, we showed that for a reaction at constant temperature, the standard Gibbs function change, $\Delta G_m^\ominus$, is directly related to the equilibrium constant. Values of $\Delta G_m^\ominus$, therefore, provide us with a way of comparing equilibrium positions for the reactions of metals with particular oxidizing agents, and in the next Section, we shall illustrate the point for the case in which the oxidizing agent is $H^+(aq)$. There we embark on an approach that will eventually provide answers to questions that lie at the heart of the problem discussed in Sections 2–4. These questions are:

1 How can we compare values of $\Delta G_m^\ominus$ for the reactions of metals with a particular oxidizing agent?

2 What is meant by the 'reactivity' of a metal, and how is it related to the value of $\Delta G_m^\ominus$ and to the rate of the reaction concerned?

3 Do values of $\Delta G_m^\ominus$ for the reactions of metals with one particular oxidizing agent remain in the same order when the oxidizing agent is changed?

4 What atomic properties cause one metal, such as sodium, to be a better reducing agent than another, such as silver (Figure 12.1)?

(a)

(b)

Figure 12.1
The strength of sodium (a) as a reducing agent becomes apparent when water is dropped on it; the water is reduced with such vigour that the remaining sodium catches fire. The weakness of silver (b) as a reducing agent is revealed by its endurance in works of art such as the Entemena vase of ancient Sumeria, dated 2500 BC.

12.1 Metals and their aqueous ions

In Sections 2–4 we asked whether we could grade metals in the order of their tendency to be oxidized, and in Section 2.2 we were especially interested in their oxidation to *aqueous* ions in reactions such as

$$Mg(s) + 2H^+(aq) = Mg^{2+}(aq) + H_2(g) \qquad (2.2)$$

In Section 4.1, we set out with the intention of comparing different metals in terms of the position of equilibrium in instances such as Reaction 2.2. Several problems had to be overcome.

Firstly, the equilibrium position in Reaction 2.2 lies well over on the right-hand side of the equation (Figure 12.2 left), and it is impossible to obtain a value of the equilibrium constant by measuring the concentrations of the species in the equation at equilibrium.

Secondly, we saw in Section 2.2 that there were reactions similar to Reaction 2.2 that did not occur, and it seemed impossible to compare the equilibrium positions for these reactions; for example, silver and dilute acid do not react (Figure 12.2 right)

$$Ag(s) + H^+(aq) = Ag^+(aq) + \tfrac{1}{2}H_2(g) \qquad (2.10)$$

The standard Gibbs function change, $\Delta G_m^{\ominus}$, goes a long way towards providing a solution to these two difficulties. From Equations 11.7 and 11.8, we know that, at constant temperature, the more negative is $\Delta G_m^{\ominus}$, the larger is the equilibrium constant of a reaction. What is more, the values of $\Delta G_m^{\ominus}$ do not necessarily have to be

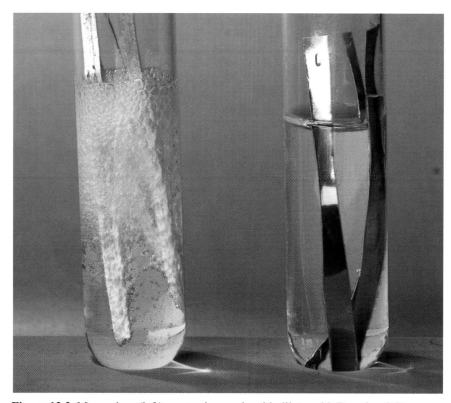

Figure 12.2 Magnesium (left) reacts vigorously with dilute acid (Reaction 2.2); silver (right) does not.

determined by studying the reaction itself; they can be calculated from the results of experiments on the individual reactants and products (usually obtainable from tables of published data).

Values of $\Delta G_m^{\ominus}$ for most reactions are calculated from standard enthalpy and entropy data, because this method has the widest application.

● For Reaction 2.2 at 298.15 K: $\Delta H_m^{\ominus} = -466.9 \, \text{kJ mol}^{-1}$; $\Delta S_m^{\ominus} = -40.1 \, \text{J K}^{-1} \text{mol}^{-1}$. What is the value of $\Delta G_m^{\ominus}$ at 298.15 K?

● The answer is $-454.9 \, \text{kJ mol}^{-1}$. The figure is obtained by substituting appropriate values in the equation

$$\Delta G_m^{\ominus} = \Delta H_m^{\ominus} - T \Delta S_m^{\ominus} \tag{11.4}$$

(Remember that $T = 298.15$ K and that $\Delta S_m^{\ominus}$ was quoted in the unit $\text{J K}^{-1} \text{mol}^{-1}$, *not* $\text{kJ K}^{-1} \text{mol}^{-1}$).

● Repeat the calculation for Equation 2.10, given that $\Delta H_m^{\ominus} = 105.6 \, \text{kJ mol}^{-1}$ and $\Delta S_m^{\ominus} = 95.5 \, \text{J K}^{-1}$.

● This time, $\Delta G_m^{\ominus} = +77.1 \, \text{kJ mol}^{-1}$.

Note how the figures concur with the observations made on Reactions 2.2 and 2.10 in Section 2.2. The value for the magnesium reaction is large and negative, so, thermodynamically, the reaction is very favourable. This is in accord with our observation that magnesium dissolves in acids to give hydrogen gas and $Mg^{2+}(aq)$ ions. The value for the silver reaction is large and positive, so the reaction is thermodynamically unfavourable and the equilibrium position lies well over on the left-hand side. This is in accord with our observation that silver does *not* dissolve in dilute acids to give hydrogen gas and $Ag^+(aq)$ ions.

Let us now move on to the question of whether the two values we have given for $\Delta G_m^{\ominus}$ for Reactions 2.2 and 2.10 ($-454.9 \, \text{kJ mol}^{-1}$ and $+77.1 \, \text{kJ mol}^{-1}$, respectively) provide a quantitative basis for comparing positions of equilibrium. Each value refers to a particular amount of material — for *one mole of reaction for the particular Equations 2.2 and 2.10*. The most useful comparison is procured by a careful choice of the amounts of material to which the values of $\Delta G_m^{\ominus}$ refer. It can be seen that in Equation 2.2, one mole of metal reacts with two moles of hydrogen ions to produce one mole of hydrogen gas. As we shall show more clearly in a moment, the values of $\Delta G_m^{\ominus}$ are most usefully compared when they refer to reactions in which the metals react with the same number of moles of aqueous hydrogen ions to produce the same number of moles of hydrogen gas.

● Convert Reaction 2.2 (and its $\Delta G_m^{\ominus}$ value) to one in which the number of moles of hydrogen ions is the same as in Reaction 2.10.

● One mole of hydrogen ions is needed on the left-hand side and half a mole of hydrogen gas on the right, so the magnesium reaction and its $\Delta G_m^{\ominus}$ value must be halved:

$$\tfrac{1}{2}\text{Mg(s)} + \text{H}^+(\text{aq}) = \tfrac{1}{2}\text{Mg}^{2+}(\text{aq}) + \tfrac{1}{2}\text{H}_2(\text{g}); \; \Delta G_m^{\ominus} = -227.5 \, \text{kJ mol}^{-1} \tag{12.1}$$

○ Subtract Equation 2.10 and its $\Delta G_m^{\ominus}$ value from Equation 12.1 and its $\Delta G_m^{\ominus}$ value.

○ All the hydrogen disappears, and you should get

$$\tfrac{1}{2}Mg(s) + Ag^+(aq) = \tfrac{1}{2}Mg^{2+}(aq) + Ag(s); \; \Delta G_m^{\ominus} = -304.6 \, kJ \, mol^{-1} \quad (12.2)$$

○ How is this last value related to the observations in Activity 2.3?

○ The equation describes the reaction of magnesium with silver ions to give silver metal and magnesium ions; the large negative value of $\Delta G_m^{\ominus}$ shows that equilibrium lies well over to the right. These conclusions are consistent with Activity 2.3, where you saw that the reaction happened.

Let us now expand this kind of operation. We can write down a series of equations for the various reactions between metals and hydrogen ions in which one mole of $H^+(aq)$ appears on the left and half a mole of hydrogen on the right. These equations will be of the general kind:

$$\tfrac{1}{n}M(s) + H^+(aq) = \tfrac{1}{n}M^{n+}(aq) + \tfrac{1}{2}H_2(g) \qquad (12.3)$$

where n is the number of positive charges on the metal cation. We can then use experimental thermodynamic data to obtain the values of $\Delta G_m^{\ominus}$ for each of these reactions, and then arrange the reactions in a table with the most negative values of $\Delta G_m^{\ominus}$ at the top and the most positive at the bottom. This has been done in Table 12.1.

If you again think carefully through the example of magnesium and silver, you will see that Table 12.1 has been arranged so that at 298.15 K any metal is thermodynamically capable of reducing the ions of another metal below it in the table to the metal itself, the first metal being simultaneously converted into its ions. This follows because $\Delta G_m^{\ominus}$ for the reaction must be negative at 298.15 K.

To convince yourself of this, try the following Question.

QUESTION 12.1

Use Table 12.1 to calculate values of $\Delta G_m^{\ominus}$ for the following reactions:

(i) $\tfrac{1}{2}Zn(s) + Ag^+(aq) = \tfrac{1}{2}Zn^{2+}(aq) + Ag(s)$ $-73.5 - 77.1 = -150.6$

(ii) $Cu(s) + Mg^{2+}(aq) = Cu^{2+}(aq) + Mg(s)$ $32.7 - (-227.5) \times 2 = 520.4$

(iii) Fc(s) $+ 2Ag^+(aq) = Fe^{2+}(aq) + 2Ag(s)$ $-39.5 - 77.1 \times 2 = -233.2$

(iv) $3Hg(l) + 2Al^{3+}(aq) = 3Hg^{2+}(aq) + 2Al(s)$ $82.2 - (-161.7) \times 6 = 1463.4$

The series of reactions for the metals in Table 12.1 is often called the *electrochemical series*. It gives us a means of comparing equilibrium constants for the reactions of metals with the oxidizing agent $H^+(aq)$. If you read other textbooks, you will often find it graded in terms of another quantity called 'the redox potential, $E^{\ominus}$', rather than in values of $\Delta G_m^{\ominus}$. You will meet this alternative later in Section 21.

The series in Table 12.1 supplies the answer to the first of our introductory questions in the particular case where the oxidizing agent is $H^+(aq)$: from thermodynamic considerations alone, the most readily oxidized metals are at the top of the series, and the least readily oxidized are at the bottom. However, as you will see in the next Section, thermodynamics alone does not determine whether a reaction occurs or not, and we must explore this point further before we tackle our second introductory question on the meaning of reactivity. At this stage, you *must* be able to manipulate

Table 12.1 Standard Gibbs function changes for the reactions of metals with aqueous hydrogen ions at 298.15 K

Reaction	$\Delta G_m^{\ominus}$/kJ mol^{-1}
$Li(s) + H^+(aq) = Li^+(aq) + \frac{1}{2}H_2(g)$	−293.3
$Cs(s) + H^+(aq) = Cs^+(aq) + \frac{1}{2}H_2(g)$	−292.0
$Rb(s) + H^+(aq) = Rb^+(aq) + \frac{1}{2}H_2(g)$	−284.0
$K(s) + H^+(aq) = K^+(aq) + \frac{1}{2}H_2(g)$	−283.3
$\frac{1}{2}Ba(s) + H^+(aq) = \frac{1}{2}Ba^{2+}(aq) + \frac{1}{2}H_2(g)$	−280.4
$\frac{1}{2}Sr(s) + H^+(aq) = \frac{1}{2}Sr^{2+}(aq) + \frac{1}{2}H_2(g)$	−279.7
$\frac{1}{2}Ca(s) + H^+(aq) = \frac{1}{2}Ca^{2+}(aq) + \frac{1}{2}H_2(g)$	−276.8
$Na(s) + H^+(aq) = Na^+(aq) + \frac{1}{2}H_2(g)$	−261.9
$\frac{1}{2}Mg(s) + H^+(aq) = \frac{1}{2}Mg^{2+}(aq) + \frac{1}{2}H_2(g)$	−227.5
$\frac{1}{3}Al(s) + H^+(aq) = \frac{1}{3}Al^{3+}(aq) + \frac{1}{2}H_2(g)$	−161.7
$\frac{1}{2}Mn(s) + H^+(aq) = \frac{1}{2}Mn^{2+}(aq) + \frac{1}{2}H_2(g)$	−114.0
$\frac{1}{2}Zn(s) + H^+(aq) = \frac{1}{2}Zn^{2+}(aq) + \frac{1}{2}H_2(g)$	−73.5
$\frac{1}{2}Fe(s) + H^+(aq) = \frac{1}{2}Fe^{2+}(aq) + \frac{1}{2}H_2(g)$	−39.5
$\frac{1}{2}Cd(s) + H^+(aq) = \frac{1}{2}Cd^{2+}(aq) + \frac{1}{2}H_2(g)$	−38.9
$\frac{1}{2}Sn(s) + H^+(aq) = \frac{1}{2}Sn^{2+}(aq) + \frac{1}{2}H_2(g)$	−13.6
$\frac{1}{2}Pb(s) + H^+(aq) = \frac{1}{2}Pb^{2+}(aq) + \frac{1}{2}H_2(g)$	−12.2
$\frac{1}{2}H_2(g) + H^+(aq) = H^+(aq) + \frac{1}{2}H_2(g)$	0
$\frac{1}{2}Cu(s) + H^+(aq) = \frac{1}{2}Cu^{2+}(aq) + \frac{1}{2}H_2(g)$	32.7
$Ag(s) + H^+(aq) = Ag^+(aq) + \frac{1}{2}H_2(g)$	77.1
$\frac{1}{2}Hg(l) + H^+(aq) = \frac{1}{2}Hg^{2+}(aq) + \frac{1}{2}H_2(g)$	82.2
$\frac{1}{3}Au(s) + H^+(aq) = \frac{1}{3}Au^{3+}(aq) + \frac{1}{2}H_2(g)$	143.7

the values of $\Delta G_f^{\ominus}$, $\Delta H_f^{\ominus}$ and $S^{\ominus}$ for the aqueous cations recorded in the *Data Book*. The procedure was covered in Question 11.5. Remember that the values of $\Delta G_f^{\ominus}$ and $\Delta H_f^{\ominus}$ are the values of $\Delta G_m^{\ominus}$ and $\Delta H_m^{\ominus}$ for reactions of the type:

$$M(s) + nH^+(aq) = M^{n+}(aq) + \frac{n}{2}H_2(g) \tag{12.4}$$

You have reached an excellent level of performance if you can do Question 12.1 above and Questions 12.2 and 12.3 below!

12.2 Summary of Section 12

Metals that form monatomic aqueous cations can be arranged in order of the values of $\Delta G_m^{\ominus}$ for the reaction

$$\frac{1}{n}M(s) + H^+(aq) = \frac{1}{n}M^{n+}(aq) + \frac{1}{2}H_2(g) \tag{12.3}$$

The tendency for the metal to be oxidized increases as ΔG_m^{O} becomes more negative.

QUESTION 12.2

This question is concerned with Table 12.2. Lanthanum (Figure 12.3a) is the first member of the lanthanide series, and plutonium (Figure 12.3b and c) is an actinide metal whose isotopes are used in nuclear reactors and nuclear weapons. Both metals dissolve in dilute acids to give hydrogen and M^{3+} ions.

Table 12.2 Thermodynamic data at 298.15 K for lanthanum and plutonium

Substance	State	$\dfrac{\Delta H_f^{\ominus}}{\text{kJ mol}^{-1}}$	$\dfrac{\Delta G_f^{\ominus}}{\text{kJ mol}^{-1}}$	$\dfrac{S^{\ominus}}{\text{J K}^{-1}\,\text{mol}^{-1}}$
H_2	g	0	0	130.7
H^+	aq	0	0	0
La	s	0	0	56.5
La^{3+}	aq	−709.0	−683.9	−223.7
Pu	s	0	0	50.3
Pu^{3+}	aq	−593.0	−585.3	−171.7

An investigator determines the values of $\Delta H_m^{\ominus}$ and $\Delta S_m^{\ominus}$ at 298.15 K for the reactions

$$La(s) + 3H^+(aq) = La^{3+}(aq) + \frac{3}{2}H_2(g) \tag{12.5}$$

$$Pu(s) + 3H^+(aq) = Pu^{3+}(aq) + \frac{3}{2}H_2(g) \tag{12.6}$$

For the lanthanum reaction, the values found are

$\Delta H_m^{\ominus} = -709.0\,\text{kJ mol}^{-1}$ and $\Delta S_m^{\ominus} = -84.1\,\text{J K}^{-1}\,\text{mol}^{-1}$

For the plutonium reaction, the values found are

$\Delta H_m^{\ominus} = -593.0\,\text{kJ mol}^{-1}$ and $\Delta S_m^{\ominus} = -25.9\,\text{J K}^{-1}\,\text{mol}^{-1}$

Fill in the six blank entries in Table 12.2.

QUESTION 12.3

What positions would lanthanum and plutonium occupy in Table 12.1?

(a)

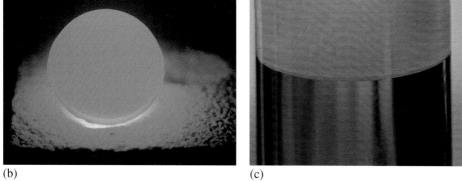

(b)   (c)

Figure 12.3 Samples of lanthanum, (a), and plutonium, (b), the two metals that are the subject of Questions 12.2 and 12.3. The photograph of plutonium has been taken by the light produced by its radioactive decay. In (c), the lower blue layer shows the colour of the aqueous plutonium ion, Pu^{3+}, which also features. It is undergoing separation from a yellow solution of uranium in the upper (kerosene) layer.

THERMODYNAMIC AND KINETIC STABILITY

13

You know that two separate conditions must be fulfilled before a reaction can occur: first, the equilibrium position must favour the reaction, and second, the reaction must have a measurable rate. The thermodynamics introduced in Sections 5–11 allows us to restate the first of these two conditions: we can now say that a reaction will occur only if $\Delta G_m^\ominus$ is negative, and if the reaction has a measurable rate.

Clearly, if some reaction does *not* occur, it could be due to a failure on either of these two fronts. Firstly, $\Delta G_m^\ominus$ could be positive: in such a case, we say that the reaction is thermodynamically unfavourable and that the reactants are **thermodynamically stable** with respect to the products. Alternatively, $\Delta G_m^\ominus$ could be negative. If this is so, then the reaction is thermodynamically favourable, and the absence of an observable reaction must be due to an immeasurably slow rate. In such a case, the reactants are said to be **kinetically stable** with respect to the reaction in question. Faced with a reaction that does not happen, you should now be able to decide which of these two kinds of stability is responsible.

To illustrate this, we shall use the reactions

$$\frac{1}{2}N_2(g) + \frac{3}{2}H_2(g) = NH_3(g) \tag{13.1}$$

$$Cu(s) + 2H^+(aq) = Cu^{2+}(aq) + H_2(g) \tag{2.9}$$

At 298.15 K, neither of these two reactions occurs. When nitrogen and hydrogen are mixed at this temperature, nothing happens: no ammonia is formed. Likewise, as you saw in Activity 2.1, copper does not react with dilute hydrochloric acid.

Use the *Data Book* and Table 12.1 to find out whether the reactants in Equations 13.1 and 2.9 are thermodynamically or kinetically stable with respect to the products.

In Reaction 13.1, ammonia is formed from its constituent elements, and $\Delta G_m^\ominus$ is equal to $\Delta G_f^\ominus(NH_3, g)$. From the *Data Book*, $\Delta G_f^\ominus(NH_3, g) = -16.5\ kJ\ mol^{-1}$.

Because $\Delta G_m^\ominus$ is negative, the equilibrium position in Equation 13.1 lies well over to the right. This means that the reaction does not occur because it is very slow: at 298.15 K, nitrogen and hydrogen are *kinetically* stable with respect to ammonia.

For Reaction 2.9, we see from Table 12.1 that $\Delta G_f^\ominus = 2 \times 32.7\ kJ\ mol^{-1} = 65.4\ kJ\ mol^{-1}$. $\Delta G_m^\ominus$ is positive, so equilibrium lies to the left of Equation 2.9: at 298.15 K, copper and $H^+(aq)$ are thermodynamically stable with respect to $Cu^{2+}(aq)$ and hydrogen gas.

The analysis of stability in terms of thermodynamics and kinetics is most important. When we want to find out why a reaction does not occur, this analysis is the first step in any rigorous argument. The procedure used for Reactions 13.1 and 2.9 is quite general: it is shown diagrammatically in Figure 13.1.

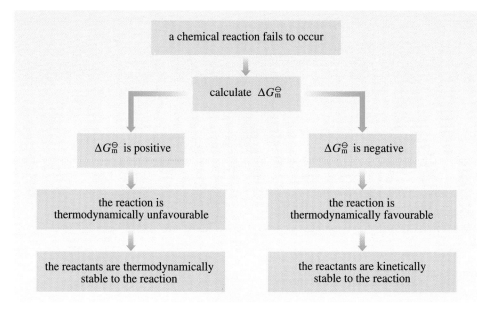

Figure 13.1 Analysis of the reasons why a reaction fails to occur.

The thermodynamics of a reaction depends only on the initial and final states of the reactants and products: thermodynamics tells us nothing about the intermediate steps, the processes that occur as the reactants are converted into the products. These processes, which constitute the *mechanism* of the reaction, are the province of reaction kinetics[*]. In the case of Reaction 13.1, for example, the mechanism is the process by which the very strong $N \equiv N$ and $H—H$ bonds in the N_2 and H_2 molecules are broken, and $N—H$ bonds are formed. Somewhere in this mechanism lies the reason for the kinetic stability of a mixture of nitrogen and hydrogen at 298.15 K.

When, as in the case of Reaction 13.1, a thermodynamically favourable reaction is immeasurably slow, we can take steps to increase the rate.

● What steps are taken in the case of Reaction 13.1?

● Reaction 13.1 is the Haber process for the preparation of ammonia. As you know, the reaction can be speeded up by using a catalyst — in this case iron — and by raising the temperature to about 750 K.

The introduction of a catalyst and an increase in temperature are quite common methods of speeding up a reaction. Other measures may be appropriate in particular cases. Consider, for example, the statue of Eros (Figure 13.2). This was one of the earliest pieces of aluminium casting: it has been standing in Piccadilly Circus since 1893. Now, $\Delta G_f^{\ominus}(Al_2O_3, s) = -1\,582.3 \text{ kJ mol}^{-1}$.

● Is the survival of Eros due to its kinetic or thermodynamic stability?

● Kinetic stability: the value of $\Delta G_f^{\ominus}$ shows that the combination of the metal with atmospheric oxygen is thermodynamically favourable. The lack of corrosion is caused by the slow rate of reaction.

* Chemical kinetics is the subject area of *Chemical Kinetics and Mechanism*[1].

Figure 13.2
The statue of Eros in Piccadilly Circus is one of the earliest examples of aluminium casting.

This slow rate has an interesting cause: when a clean piece of aluminium foil is exposed to oxygen, a thin film of Al_2O_3, about 10^{-8} m thick, is quickly formed on the metal surface. This film prevents further contact between the metal and oxygen so that the reaction then ceases: here the mechanism of the reaction incorporates a step that stops the reaction dead!

One way of speeding up the reaction is to bring the reactants into more intimate contact by grinding the aluminium into a very fine powder. This increases the surface to volume ratio so much that oxide formation does not stop with a surface layer: very fine aluminium powders can actually explode spontaneously in air.

A more subtle method is shown in Activity 4.1 (on the CD-ROM associated with this Book): a little mercury or mercury(II) chloride will destroy the coherence of the oxide film, and the underlying aluminium dissolves in the mercury to form an alloy called an *amalgam*. The aluminium in the amalgam will then combine quickly with atmospheric oxygen to form white flakes of oxide (Figure 13.3).

Figure 13.3 Unlike untreated aluminium (left), aluminium whose surface has been briefly amalgamated by contact with mercury dichloride solution reacts vigorously with atmospheric oxygen or water.

The factors that affect the kinetic stability of aluminium at 298.15 K are clearly very different from those responsible for the kinetic stability of a mixture of nitrogen and hydrogen at the same temperature. They show how wide-ranging in scope such factors can be: *kinetic stability can be attributed to anything that prevents a thermodynamically favourable reaction from taking place.*

13.1 Summary of Section 13

1   If $\Delta G_m^{\ominus}$ for a reaction is positive at any temperature, the reactants are thermodynamically stable with respect to the products: the reaction will never occur
at that temperature.

2   If $\Delta G_m^{\ominus}$ for a reaction is negative, the reaction may occur or it may not. It will
occur if the rate of reaction is sufficiently high. It will not occur if the rate of
reaction is immeasurably slow — that is, if the reactants are kinetically stable
with respect to the products.

3   An immeasurably slow but thermodynamically favourable reaction can possibly
be speeded up by finding a catalyst, or by bringing the reactants into more
intimate contact, or by increasing the temperature.

QUESTION 13.1

Ignoring the possibility of slow rates of reaction, which of the four reactions in
Question 12.1 would you expect to occur?

QUESTION 13.2

In Activities 2.3 and 4.1, you saw that the following reactions did not occur:

$Cu(s) + Mg^{2+}(aq) = Cu^{2+}(aq) + Mg(s)$

$2Al(s) + 3Cu^{2+}(aq) = 2Al^{3+}(aq) + 3Cu(s)$

Decide in each case whether the lack of reaction is due to kinetic or
thermodynamic stability.

QUESTION 13.3

In proving that a stable chemical system is kinetically rather than
thermodynamically stable, one must find a reaction that the system might
undergo and show that it has a negative value of $\Delta G_m^{\ominus}$. Do this for the
following systems in which no reaction occurs at 298.15 K:

(a) a mixture of diamonds and oxygen at 298.15 K;

(b) a mixture of magnesium ribbon and solid zinc chloride ($ZnCl_2$) at 298.15 K.

REACTIVITY

14

We have now arrived at the point where we can begin to answer the second of the three introductory questions in Section 12. This was concerned with the meaning of reactivity.

In Sections 2–4, we used merely visual observation of reactions to grade metals in order of their reactivity or tendency to be oxidized. Let us look at a typical example of the style of reasoning that we used:

> 'Suppose we take three beakers of dilute hydrochloric acid, and drop a piece of silver into the first, lithium into the second and potassium into the third. In the first case, nothing happens. In the second, the lithium floats steadily around on the surface evolving hydrogen. In the third case, the potassium streaks around the acid surface, and the evolved hydrogen catches fire (see Figure 14.1). Clearly potassium is more reactive than lithium towards dilute acid, and lithium is more reactive than silver.'

This kind of reasoning is compelling because it is directly derived from the full drama of what we see. We shall now show how it conflicts with the logic of Section 13. Table 14.1 lists values of $\Delta G_m^{\ominus}$ and $\Delta H_m^{\ominus}$ for the appropriate reactions of the three metals.

Table 14.1 Thermodynamic data for the reactions of lithium, potassium and silver with dilute acid

Reaction	$\Delta G_m^{\ominus}$/kJ mol^{-1}	$\Delta H_m^{\ominus}$/kJ mol^{-1}
$Li(s) + H^+(aq) = Li^+(aq) + \frac{1}{2}H_2(g)$	−293.3	−278.5
$K(s) + H^+(aq) = K^+(aq) + \frac{1}{2}H_2(g)$	−283.3	−252.4
$Ag(s) + H^+(aq) = Ag^+(aq) + \frac{1}{2}H_2(g)$	77.1	105.6

(a)

(b)

(c)

Figure 14.1
The reactions of (a) silver, (b) lithium and (c) potassium with dilute acid.

⬤ Are the differences in reactivity attributable to a kinetic or thermodynamic effect?

⬤ The silver reaction is distinguished from the other two by a positive value of $\Delta G_m^{\ominus}$: it is thermodynamically unfavourable. The difference between silver on the one hand and lithium or potassium on the other is due to thermodynamic effects. But the difference between lithium and potassium is much more complicated.

The values of $\Delta G_m^{\ominus}$ show that equilibrium is further to the right for lithium, and the values of $\Delta H_m^{\ominus}$ show that more heat is evolved per mole of metal in the lithium reaction. The reason why potassium is more reactive must be that its reaction with hydrogen ions is faster. Heat is generated quickly in the potassium reaction, so there is a localized increase of temperature around the dissolving metal, which causes the hydrogen to catch fire.

The symptoms by which we recognize reactivity, such as marked rise in temperature and violent agitation of the reactants and products, arise from a peculiar combination of kinetic and thermodynamic effects. We can distinguish two quite basic requirements, and a third which is less important.

1   One obvious requirement that must be fulfilled before a chemical system can be described as reactive is that the reaction should be very fast — a question of kinetics.

2   Equally obvious is the requirement that the reaction should be thermodynamically favourable — a question of thermodynamics that is contained in the condition that $\Delta G_m^{\ominus}$ must be negative.

3   Visual impressions of reactivity are enhanced if $\Delta H_m^{\ominus}$ is large and negative, because if the initial reaction is fast, heat will be evolved quickly, leading to a localized rise in temperature. This rise in temperature increases the rate still further, and the symptoms of reactivity become even more marked.

To summarize, the observed 'reactivity' is an ill-defined union of thermodynamic and kinetic effects. 'Reactive' materials usually display their reactivity in reactions that not only have large equilibrium constants and large negative values of $\Delta H_m^{\ominus}$, but also proceed rapidly. Two conclusions follow from this.

Firstly, when you say that a substance is reactive, try to specify those reactions through which its reactivity is expressed. In other words, compare the reactivity of reactions rather than that of individual substances. When you have done this, analyse the behaviour in terms of thermodynamics and kinetics.

Secondly, since quantitative scientific accuracy demands that we separate kinetic and thermodynamic effects, we cannot use 'reactivity' to compare the reactions of metals. We must make either kinetic or thermodynamic comparisons. Because we can compare the thermodynamics but not the rates in cases where reactions do not occur, we shall in fact use thermodynamics to make the comparison. It is on these terms that we shall tackle the third of our three introductory questions.

THERMODYNAMICS AND THE OXIDATION OF METALS

15

In this Section, we shall use thermodynamics to explore the extent to which we can grade metals in order of their ease of oxidation.

The experiments and observations of Sections 2–4 suggested that it might be possible to produce a single grading of metals arranged in order of their tendency to be oxidized by various oxidizing agents. As we said at the time, this idea was prominent in the nineteenth century, and it led to quite detailed but essentially qualitative gradings, of which the following example is typical:

Cs > Rb > K > Na > Li > Ba > Sr > Ca > Mg
> Al > Zn > Fe > Sn > Pb > H > Cu > Hg > Ag > Au

Now, in Section 12.1 we used thermodynamic data to set up a *quantitative* grading of metals in order of their thermodynamic willingness to be oxidized by aqueous hydrogen ions.

⬤ Compare the grading given immediately above with that in Table 12.1. Do the two correspond (a) in broad outline, (b) in detail?

⬤ They certainly correspond in broad outline. For example, in each case the alkali and alkaline earth metals come first, whereas copper, silver and gold are last. Detailed study shows that from magnesium onwards, the two series are almost identical, but there are serious discrepancies among the first eight metals. Compare, for example, the positions of lithium and sodium in the two series.

Similarities of this kind suggest one of two possibilities: firstly, there might be a unique series (but either one or both of the two that we have compared gives untrustworthy results for the first eight metals); secondly, there is no unique series. We can resolve these alternatives by setting up another table in which the metals are arranged in order of their thermodynamic willingness to be oxidized by a different oxidizing agent.

In Table 12.1, we used aqueous hydrogen ions as the oxidizing agent, and wrote down a series of equations in which the same amount of $H^+(aq)$ (one mole) appeared on the left-hand side. On subtraction of the equations, hydrogen ions and hydrogen gas disappeared. But from a practical standpoint, the most important oxidizing agent to which metals are vulnerable is the molecular oxygen in the air. The oxidizing power of molecular oxygen, often augmented by the presence of water, is the source of metal corrosion, an immensely costly and wasteful process. In Table 15.1, we have used molecular oxygen as our oxidizing agent and written down the equations with the same amount of oxygen (half a mole) on the left-hand side. Again, on subtraction of the equations, oxygen gas disappears. The equations have been arranged, as in Table 12.1, with the most negative values of $\Delta G_m^\ominus$ at the top. (The values are taken from the *Data Book*.)

Table 15.1 Values of $\Delta G_m^{\ominus}$ for the reactions of different metals with molecular oxygen at 298.15 K

Reaction	$\Delta G_m^{\ominus}$/kJ mol^{-1}	Reaction	$\Delta G_m^{\ominus}$/kJ mol^{-1}
$Ca(s) + \frac{1}{2}O_2(g) = CaO(s)$	−604.0	$2Rb(s) + \frac{1}{2}O_2(g) = Rb_2O(s)$	−290.8
$Mg(s) + \frac{1}{2}O_2(g) = MgO(s)$	−569.4	$\frac{1}{2}Sn(s) + \frac{1}{2}O_2(g) = \frac{1}{2}SnO_2(s)$	−259.8
$Sr(s) + \frac{1}{2}O_2(g) = SrO(s)$	−561.9	$Fe(s) + \frac{1}{2}O_2(g) = FeO(s)$	−251.4
$2Li(s) + \frac{1}{2}O_2(g) = Li_2O(s)$	−561.2	$H_2(g) + \frac{1}{2}O_2(g) = H_2O(l)$	−237.1
$\frac{2}{3}Al(s) + \frac{1}{2}O_2(g) = \frac{1}{3}Al_2O_3(s)$	−527.4	$Pb(s) + \frac{1}{2}O_2(g) = PbO(s)$	−188.9
$Ba(s) + \frac{1}{2}O_2(g) = BaO(s)$	−525.1	$Cu(s) + \frac{1}{2}O_2(g) = CuO(s)$	−129.7
$2Na(s) + \frac{1}{2}O_2(g) = Na_2O(s)$	−375.5	$Hg(l) + \frac{1}{2}O_2(g) = HgO(s)$	−58.5
$2K(s) + \frac{1}{2}O_2(g) = K_2O(s)$	−320.7	$2Ag(s) + \frac{1}{2}O_2(g) = Ag_2O(s)$	−11.2
$Zn(s) + \frac{1}{2}O_2(g) = ZnO(s)$	−318.3	$\frac{2}{3}Au(s) + \frac{1}{2}O_2(g) = \frac{1}{3}Au_2O_3(s)$	54.4
$2Cs(s) + \frac{1}{2}O_2(g) = Cs_2O(s)$	−308.1		

You can, of course, manipulate the equations and $\Delta G_m^{\ominus}$ values in Table 15.1 in exactly the same way as you did with the ones in Table 12.1. This exercise forms part of Question 15.1. However, here we concentrate on the comparison between the order of the metals in Tables 12.1 and 15.1.

Compare the order of the metals in Tables 12.1 and 15.1. Do the two series correspond (a) in broad outline, (b) in detail?

The similarities are not as close as last time, but they are still apparent. Again the alkali and alkaline earth metals come at the top, whereas copper, silver and gold are at the bottom. This time, the detailed differences are more striking: compare, for example, the positions of magnesium and aluminium.

Which of the two possibilities mentioned earlier in this Section is closer to the truth?

The second; the differences between the two sequences in Tables 12.1 and 15.1 suggest that the concept of a unique series is an illusion. No single series can express the thermodynamic tendency of metals to be oxidized by different oxidizing agents, and the goal of a unique series, whose possible existence was suggested in Sections 2.3 and 4, is unattainable.

However, just as important is the fact that the series do have overall similarities: the alkali and alkaline earth metals are near the top, whereas silver and gold are at the bottom. We shall return to this point later in the Book.

15.1 Some consequences of Table 15.1

We start this Section with a question:

● For how many of the metals in Table 15.1 is the value of ΔG negative?

● For all except gold.

This is very remarkable. Only for gold is $\Delta G_f^{\ominus}$ (Au_2O_3, s) positive. In all the other cases, oxidation of the metal by atmospheric oxygen is thermodynamically favourable at 25 °C. Because we rely so heavily on metals that exist in contact with atmospheric oxygen, our civilization depends on the fact that the oxidation of all metals in common use is slow. Only kinetic barriers stand in the way of the structural collapse of many of life's essentials, including cutlery, the car and the jumbo jet.

Another feature of Table 15.1 is that for most of the metals in it, the values of $\Delta G_m^{\ominus}$ are very negative. Now from Equation 11.4:

$$\Delta G_m^{\ominus} = \Delta H_m^{\ominus} - T\Delta S_m^{\ominus} \tag{11.4}$$

So the negative values arise either from very negative values of $\Delta H_m^{\ominus}$ or very positive figures for $\Delta S_m^{\ominus}$. We shall have more to say about this matter in Section 16. However, the essential point is that very negative values of $\Delta H_m^{\ominus}$ are usually responsible. Consequently, the oxidation of many of the metals in Table 15.1 is associated with the evolution of large amounts of heat. In Section 14 we noted that this makes it likely that, once started, the reaction will proceed at an increasing and perhaps catastrophic rate.

The case of aluminium shows how tenacious oxide films are often the reason why vigorous reaction does not occur. In Section 14, you saw that corrosion is rapid if the film is disrupted by amalgamation. Small particle sizes, at which the surface-to-volume ratio is much larger than for bulk aluminium, also threaten the coherence of the film. In workplaces, where aluminium powder is handled, or aluminium dust circulates, a small spark can ignite the metal and cause accidents. Not surprisingly then, aluminium powder is used in fireworks and flares to produce an intense white light. The next Section describes how a similar reaction has been invoked in an explanation of the strange phenomenon of ball lightning.

15.1.1 Ball lightning

Everyone is familiar with the common type of lightning, shown in Figure 15.1a as a jagged discharge or 'bolt'. **Ball lightning**, as described in Box 15.1 by the American physicist, Graham Hubler, is very different. It consists of a glowing sphere of light, about the size of a grapefruit, which usually appears out of thin air during thundery weather. The spheres float close to the ground, drifting slowly for about 20 seconds before expiring. Recorded sightings of ball lightning go back to the Middle Ages, although its occurrence is very rarely observed. A few photographs exist (Figure 15.1b), but as yet [2002], there are no known videotapes. Until recently, there was no convincing scientific explanation; it sounds, in fact, like the stuff of which UFOs are made.

BOX 15.1 Ball lightning seen and described

I saw ball lightning during a thunderstorm in the summer of 1960. I was 16 years old. It was about 9pm, very dark, and I was sitting with my girlfriend at a picnic table in a pavilion at a public park in upstate New York. The structure was open on three sides and we were sitting with our backs to the closed side. It was raining quite hard. A whitish-yellowish ball, about the size of a tennis ball, appeared on our left, 30 yards away, and its appearance was not directly associated with a lightning strike. The wind was light. The ball was eight feet off the ground and drifting slowly toward the pavilion. As it entered, it dropped abruptly to the wet wood plank floor, passing within three feet of our heads on the way down. It skittered along the floor with a jerky motion (stick-slip), passed out of the structure on the right, rose to a height of six feet, drifted ten yards further, dropped to the ground and extinguished non-explosively. As it passed my head, I felt no heat. The acoustic emission I liken to that of a freshly struck match. As it skittered on the floor it displayed elastic properties… Its luminosity was such that it was not blinding. I estimate it was like staring at a less than 10-watt light bulb. The whole encounter lasted for about 15 seconds. I remember it vividly even today, as all eyewitnesses do, because it was so extraordinary. Not until 10 years later, at a seminar on ball lightning, did I realize what I had witnessed.

Graham Hubler

Figure 15.1 (a) The familiar but spectacular lightning bolt. (b) A photograph of ball lightning taken in Austria in 1978. It has a white centre that extends into a luminous tail, and there is a blue surround. This makes it even more striking than the ordinary spherical sort that prompted the explanation given in the text.

A recent explanation of ball lightning depends on properties of the semi-metal silicon. The combination of silicon bound to oxygen is the principal component of soils and rocks; sand, for example, is mostly the oxide of silicon, SiO_2. The reaction between silicon and oxygen is thermodynamically very favourable:

$$\tfrac{1}{2} Si(s) + \tfrac{1}{2} O_2(g) = \tfrac{1}{2} SiO_2(s); \ \Delta G_m^{\ominus} = -428.3 \text{ kJ mol}^{-1} \qquad (15.1)$$

⬤ Where does this value place silicon in Table 15.1?

⬤ Beneath barium and above sodium.

The thermodynamic favourability of the oxidation of silicon is therefore comparable with that of highly reactive alkali and alkaline earth metals. It is also associated with the evolution of much heat: for Reaction 15.1, $\Delta H_m^{\ominus} = -455.5 \text{ kJ mol}^{-1}$. So if hot silicon is created in a finely divided state, we can expect a vigorous exothermic reaction in the presence of oxygen.

In 2000, two chemical engineers in New Zealand, John Abrahamson and James Dinniss, used this idea to produce a new and promising explanation of ball lightning. A conventional lightning strike on soil can generate temperatures as high as 3 000 K. Silicates or silica could then react with carbon from the organic matter in the soil to produce a vapour containing silicon atoms. This vapour is blown into the air by the shock wave of the lightning strike. As the silicon atoms cool, they combine to form a buoyant fluff ball composed of chains of tiny particles of solid silicon. This fluff ball burns steadily in air as the finely divided silicon combines with oxygen to regenerate SiO_2. When the silicon has been consumed, the glowing sphere disappears.

Abrahamson and Dinniss exposed soil to a lightning-like discharge in the laboratory. Although they did not see ball lightning, they did find the required chains of tiny silicon particles in the air space close to the strike. This is promising. Experiments are under way to find the soil and discharge conditions that might generate ball lightning.

15.2 Summary of Sections 14 and 15

1   'Reactive' materials usually reveal their reactivity by reactions that not only have large negative values of $\Delta H_m^{\ominus}$ and $\Delta G_m^{\ominus}$, but are also fast. Thus, 'reactivity', as usually understood, is a combination of thermodynamic and kinetic factors.

2   There is no single order of metals that expresses their thermodynamic tendency to be oxidized; different oxidizing agents yield different orders. However, some metals, such as the alkali metals, are invariably more readily oxidized than others, such as copper, silver and gold.

QUESTION 15.1

Use the data in Table 15.1 to calculate values of $\Delta G_m^{\ominus}$ for the following reactions at 298.15 K:

(i) $2Al(s) + 3PbO(s) = Al_2O_3(s) + 3Pb(s)$

(ii) $Cu(s) + Li_2O(s) = CuO(s) + 2Li(s)$

QUESTION 15.2

Use information from the *Data Book* to construct a table similar to Tables 12.1 and 15.1, representing the thermodynamic tendency of the metals Ag, Al, Ca, Cu, Hg, Na and Zn to be oxidized by chlorine at 298.15 K. Does the table support the conclusions reached in Section 15?

QUESTION 15.3

Reactions obtained by subtracting equations in Table 12.1, and predicted to be thermodynamically favourable, very often occur at 298.15 K. By contrast, reactions obtained by subtracting equations in Table 15.1 and predicted to be thermodynamically favourable, very often do not occur at 298.15 K. Suggest a reason for this difference.

ENTHALPY AND ENTROPY TERMS

16

Before moving on to the next stage of our investigation of the redox reactions of metals, we must look a little more closely at the enthalpy and entropy components of the Gibbs function. The important equation is

$$\Delta G_m^{\ominus} = \Delta H_m^{\ominus} - T\Delta S_m^{\ominus} \tag{11.4}$$

A reaction is thermodynamically favourable if $\Delta G_m^{\ominus}$ is negative. According to Equation 11.4, this situation is furthered by negative values of the standard molar enthalpy change, $\Delta H_m^{\ominus}$, and positive values of the standard molar entropy change, $\Delta S_m^{\ominus}$. Conversely, positive values of $\Delta H_m^{\ominus}$ and negative values of $\Delta S_m^{\ominus}$ tend to make reactions thermodynamically unfavourable. The sign of $\Delta G_m^{\ominus}$ is determined by the balance between the enthalpy and entropy terms on the right-hand side of Equation 11.4.

By surveying the thermodynamics of a wide variety of chemical reactions at normal temperature, one finds that in most cases, the enthalpy term is more influential than the entropy term in determining the sign of $\Delta G_m^{\ominus}$. Let us begin by looking at the thermodynamics of some reactions from Table 15.1. Consider the reaction

$$Zn(s) + \frac{1}{2}O_2(g) = ZnO(s) \tag{16.1}$$

at 298.15 K. Here, using the *Data Book*,

$$\Delta S_m^{\ominus} = S^{\ominus}(ZnO, s) - S^{\ominus}(Zn, s) - \frac{1}{2}S^{\ominus}(O_2, g)$$
$$= (43.6 - 41.6 - 102.6)\,J\,K^{-1}\,mol^{-1}$$
$$= -100.6\,J\,K^{-1}\,mol^{-1}$$

Now the *Data Book* also tells us that $\Delta H_m^{\ominus} = -348.3\,kJ\,mol^{-1}$. Thus

$$\Delta G_m^{\ominus} = -348.3\,kJ\,mol^{-1} - (298.15 \times -100.6)\,J\,mol^{-1}$$
$$= -348.3\,kJ\,mol^{-1} + 30.0\,kJ\,mol^{-1}$$
$$= -318.3\,kJ\,mol^{-1}$$

Notice that in this calculation, the numerical value of the $T\Delta S_m^{\ominus}$ term ($30.0\,kJ\,mol^{-1}$) is only about 9% of the numerical value of the $\Delta H_m^{\ominus}$ term ($348.3\,kJ\,mol^{-1}$). We show this diagrammatically on the left of Figure 16.1. Because $T\Delta S_m^{\ominus}$ is small compared with $\Delta H_m^{\ominus}$, $\Delta G_m^{\ominus}$ does not differ greatly from $\Delta H_m^{\ominus}$. For most reactions at room temperature, $\Delta H_m^{\ominus}$ makes a much larger contribution to $\Delta G_m^{\ominus}$ than does $T\Delta S_m^{\ominus}$. Indeed, this fact explains the plausibility of Thomsen's hypothesis, which you met earlier in the Book (see, for example, Section 6).

- In Equation 11.4 at 298.15 K, how many $J\,K^{-1}\,mol^{-1}$ in the $\Delta S_m^{\ominus}$ term are equivalent to $1\,kJ\,mol^{-1}$ in the $\Delta H_m^{\ominus}$ term?

- If $T\Delta S_m^{\ominus}$ is to make a $1\,kJ\,mol^{-1}$ contribution to $\Delta G_m^{\ominus}$ at 298.15 K, $\Delta S_m^{\ominus}$ must be $1\,000/298.15 = 3.35\,J\,K^{-1}\,mol^{-1}$. Thus, to the nearest whole number, an entropy change of $3\,J\,K^{-1}\,mol^{-1}$ is equivalent to only $1\,kJ\,mol^{-1}$.

Because $T \Delta S_{\mathrm{m}}^{\ominus}$ is often small compared with $\Delta H_{\mathrm{m}}^{\ominus}$, we shall occasionally ignore it altogether and assume that $\Delta G_{\mathrm{m}}^{\ominus}$ is equal to $\Delta H_{\mathrm{m}}^{\ominus}$. However, we shall most frequently ignore the $T \Delta S_{\mathrm{m}}^{\ominus}$ term when we are comparing the values of $\Delta G_{\mathrm{m}}^{\ominus}$ for *similar* types of reaction. To demonstrate this, we can compare the formation of zinc oxide, ZnO, with that of magnesium oxide, MgO. For the reaction

$$\mathrm{Mg(s)} + \tfrac{1}{2} \mathrm{O}_2(\mathrm{g}) = \mathrm{MgO(s)} \tag{2.5}$$

at 298.15 K,

$$\begin{aligned} \Delta S_{\mathrm{m}}^{\ominus} &= S^{\ominus}(\mathrm{MgO, s}) - S^{\ominus}(\mathrm{Mg, s}) - \tfrac{1}{2} S^{\ominus}(\mathrm{O}_2, \mathrm{g}) \\ &= (26.9 - 32.7 - 102.6)\,\mathrm{J\,K^{-1}\,mol^{-1}} \\ &= -108.4\,\mathrm{J\,K^{-1}\,mol^{-1}} \end{aligned}$$

Notice that the value of $\Delta S_{\mathrm{m}}^{\ominus}$ for Reaction 2.5 differs by only about $8\,\mathrm{J\,K^{-1}\,mol^{-1}}$ from the value for Reaction 16.1 ($-108.4\,\mathrm{J\,K^{-1}\,mol^{-1}}$ and $-100.6\,\mathrm{J\,K^{-1}\,mol^{-1}}$, respectively). As we move from the zinc to the magnesium reaction, the molar entropy of the metal decreases substantially, but so does the value for the oxide. Hence the change in the difference between them is relatively small (about $8\,\mathrm{J\,K^{-1}\,mol^{-1}}$). The large value of $\tfrac{1}{2} S^{\ominus}(\mathrm{O}_2, \mathrm{g})$ is common to both reactions, so it does not affect the difference between the values of $\Delta S_{\mathrm{m}}^{\ominus}$. Using the *Data Book*, we obtain for the magnesium reaction at 298.15 K

$$\Delta H_{\mathrm{m}}^{\ominus} = -601.7\,\mathrm{kJ\,mol^{-1}}$$

and

$$\begin{aligned} \Delta G_{\mathrm{m}}^{\ominus} &= -601.7\,\mathrm{kJ\,mol^{-1}} - (298.15 \times -108.4)\,\mathrm{J\,mol^{-1}} \\ &= -569.4\,\mathrm{kJ\,mol^{-1}} \end{aligned}$$

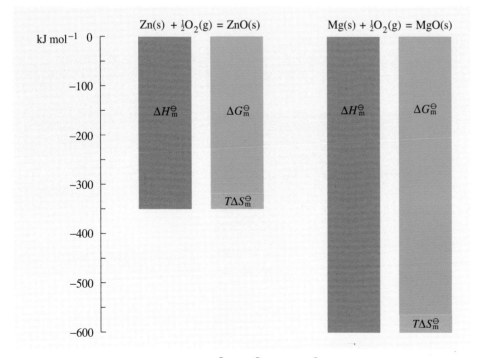

Figure 16.1 Relative magnitudes of $\Delta G_{\mathrm{m}}^{\ominus}$, $\Delta H_{\mathrm{m}}^{\ominus}$ and $T \Delta S_{\mathrm{m}}^{\ominus}$ for two reactions at 298.15 K: reaction of zinc with oxygen (left); reaction of magnesium with oxygen (right).

We can compare the two sets of calculations diagrammatically, as in Figure 16.1, or numerically. You will notice that for both reactions, the $T\Delta S_m^\ominus$ term is small compared with $\Delta H_m^\ominus$. But notice too, that the difference between the values of $\Delta G_m^\ominus$ (251.1 kJ mol^{-1}) is almost entirely due to the difference in the values of $\Delta H_m^\ominus$ (253.4 kJ mol^{-1}). This is because the two values of $\Delta S_m^\ominus$ are so similar. The result is that when we compare two values of $\Delta G_m^\ominus$ for similar types of reaction, the error incurred by ignoring the $\Delta S_m^\ominus$ term and simply comparing the $\Delta H_m^\ominus$ values is usually very small.

Before closing this Section, we shall say a little more about the meaning of the expression, 'similar types of reaction'. In the case of Reactions 16.1 and 2.5, it means that the equations differ only because one element (magnesium) has been substituted for another (zinc). We shall call reactions that are related in this way **analogous reactions**. Particularly important is the fact that the substitution does not involve any change in the numbers that are in, or that precede, chemical formulae, or in the physical states of reactants and products. Thus, solid zinc and solid zinc oxide are replaced by solid magnesium and solid magnesium oxide. The importance of this proviso should become more obvious when you have read the next Section.

16.1 Reactions of solids and gases: the sign of $\Delta S_m^\ominus$

In most instances, the form of a chemical equation reveals nothing about the sign or magnitude of $\Delta G_m^\ominus$, $\Delta H_m^\ominus$ or $\Delta S_m^\ominus$. However, from Section 10.6 you know that, when the reaction involves only pure substances, some of which are gases, the sign of $\Delta S_m^\ominus$ can usually be predicted with reasonable confidence. Consider, for example, Reactions 16.1 and 2.5. These have similar large negative values of $\Delta S_m^\ominus$.

⬤ What feature of the equation leads us to expect this?

⬤ From Section 10.5, you know that, in general, gases have much larger molar entropies than solids or liquids. Now in the solid–gas Reactions 16.1 and 2.5, there is half a mole of oxygen gas on the left-hand side of the equations, but no moles of gas on the right. The entropy of the reactants is therefore much greater than that of the products.

Thus, one obvious reason why the two analogous reactions have similar large negative entropy changes is that they both involve a decrease of half a mole of the same gas per mole of reaction. Conversely, a reaction involving solids and gases which leads to an increase in the number of moles of gas will have a large positive value of $\Delta S_m^\ominus$ (Figure 16.2).

(a)

(b)

Figure 16.2
(a) The combustion of magnesium consumes oxygen gas and has a large negative entropy change; (b) the explosion of solid nitrocellullose produces gases, and has a large positive entropy change.

16.2 Summary of Section 16

1   For the majority of chemical reactions at 298.15 K, $\Delta H_m^{\ominus}$ in the equation

$$\Delta G_m^{\ominus} = \Delta H_m^{\ominus} - T\Delta S_m^{\ominus} \qquad (11.4)$$

makes a bigger contribution to $\Delta G_m^{\ominus}$ than does $-T\Delta S_m^{\ominus}$.

2   For a series of analogous reactions at 298.15 K, the variations in $\Delta G_m^{\ominus}$ are determined almost entirely by those in $\Delta H_m^{\ominus}$ because the $T\Delta S_m^{\ominus}$ values are similar.

3   Reactions involving solids and gases have large positive values of $\Delta S_m^{\ominus}$ when there is an increase in the number of moles of gas, and large negative values of $\Delta S_m^{\ominus}$ when there is a decrease in the number of moles of gas.

QUESTION 16.1

In one of the following three pairs of reactions, the two reactions have very similar values of $\Delta S_m^{\ominus}$; which pair?

(i) $CuO(s) = Cu(s) + \frac{1}{2}O_2(g)$

   $HgO(s) = Hg(g) + \frac{1}{2}O_2(g)$

(ii) $MgCO_3(s) = MgO(s) + CO_2(g)$

   $CaCO_3(s) = CaO(s) + CO_2(g)$

(iii) $SnO_2(s) = Sn(s) + O_2(g)$

   $SnO(s) = Sn(s) + \frac{1}{2}O_2(g)$

QUESTION 16.2

From reactions (i)–(v) below, select those that (a) have large negative entropy changes, (b) have large positive entropy changes, (c) have small entropy changes that may be either positive or negative.

(i) $K(s) + \frac{1}{2}Cl_2(g) = KCl(s)$

(ii) $K(s) + \frac{1}{2}I_2(s) = KI(s)$

(iii) $C(s) + O_2(g) = CO_2(g)$

(iv) $CaCO_3(s) = CaO(s) + CO_2(g)$

(v) $Al_2O_3(s) + 3C(s) + 3Cl_2(g) = 2AlCl_3(g) + 3CO(g)$

METALS AND THEIR ORES

17

In Sections 2.4 and 3, we pointed out one of the most striking features of the chemical behaviour of metals: a particular metal often responds similarly towards quite different oxidizing agents. We indicated a relation between this observation and the details of metal extraction from ores. Thus, gold reacts very little, if at all, with the various corrosive substances found in the lithosphere and hydrosphere. Consequently, it is found naturally in the uncombined state. By contrast, the alkali and alkaline earth metals react vigorously with water, oxygen, chlorine or hydrogen chloride, and are never found free in nature. They occur as aqueous ions, chlorides, etc., which must be reduced if the metal is to be isolated. We distinguished three classes of metals:

1   'Noble' metals, which are either found free or can be obtained by gently heating their ores in air (Au, Ag, Hg; see Figure 17.1).

2   Metals that are obtained by heating their ores in air to form oxides, which can then be reduced by heating with carbon (Pb, Sn, Fe, Zn; see Figure 17.2).

Figure 17.1
The nobility of silver illustrated by the endurance of one of the most famous of all silver coins. The Athenian silver tetradrachm of the fifth century BC was the monetary symbol of the economic power of the Athenian empire; on one side is the 'almond-eyed' goddess Athena, and on the other, her symbol, the owl.

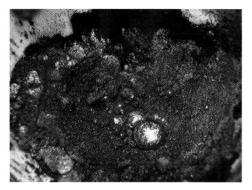

Figure 17.2
Molten lead being produced by heating carbon with orange lead monoxide (litharge or PbO).

3   Metals that are not easily obtained by carbon reduction and were isolated in relatively recent times, either with the help of electrolytic methods, or by using metals in the same class as reducing agents (alkali and alkaline earth metals, aluminium; see Figure 17.3).

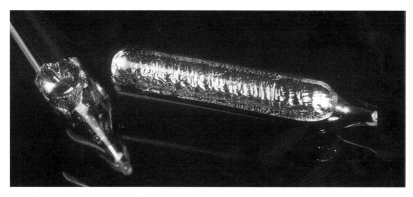

Figure 17.3
Caesium is made by heating its molten chloride with calcium, and stored in sealed ampoules. The solid is silvery, but if an ampoule is held in the hand, the caesium melts, and becomes a golden liquid.

Although rates of reaction are important in metal extraction, it is still very helpful to view this classification from a thermodynamic standpoint. We shall now do this.

17.1 The variation of $\Delta G_m^{\ominus}$ with temperature

Until now, we have handled $\Delta G_m^{\ominus}$ values at only one temperature: 298.15 K. But many reactions and industrial processes, including metal extraction, must be carried out at higher temperatures. Sometimes there is a need to speed up reactions that are kinetically unfavourable; at other times the intention is to improve the yield by increasing the equilibrium constant. But whatever the reason for the heating process, the values of the equilibrium constant and $\Delta G_m^{\ominus}$ are different at the higher temperature. This brings us up against a question of the greatest importance: how does $\Delta G_m^{\ominus}$ change with temperature?

We know that for any reaction at *any* temperature,

$$\Delta G_m^{\ominus} = \Delta H_m^{\ominus} - T \Delta S_m^{\ominus} \tag{11.4}$$

Tables like those in the *Data Book* enable us to calculate $\Delta G_m^{\ominus}$, $\Delta H_m^{\ominus}$ and $\Delta S_m^{\ominus}$ for reactions at 298.15 K.

Suppose now that the temperature is increased substantially, and that the physical states of the reactants and products remain what they were at 298.15 K; that is, no reactant or product melts or boils within the range of temperature increase. *It is then found experimentally that the values of $\Delta H_m^{\ominus}$ and $\Delta S_m^{\ominus}$ change relatively little.*

As an illustration of this fact, Figure 17.4 shows plots of $\Delta G_m^{\ominus}$, $\Delta H_m^{\ominus}$ and $\Delta S_m^{\ominus}$ against temperature for the reaction

$$Mg(s) + \tfrac{1}{2}O_2(g) = MgO(s) \tag{2.5}$$

in the range 250–900 K. Magnesium and magnesium oxide both melt above the top of this temperature range, but within it no phase changes occur. The values of $\Delta H_m^{\ominus}$ and $\Delta S_m^{\ominus}$ are remarkably constant throughout the range, but $\Delta G_m^{\ominus}$ changes by quite large amounts.

Many other examples of the same effect could be given. We shall therefore use the following approximation for all chemical reactions: provided that no reactant or product melts or boils, $\Delta H_m^{\ominus}$ and $\Delta S_m^{\ominus}$ do not change as the temperature is raised above 298.15 K.

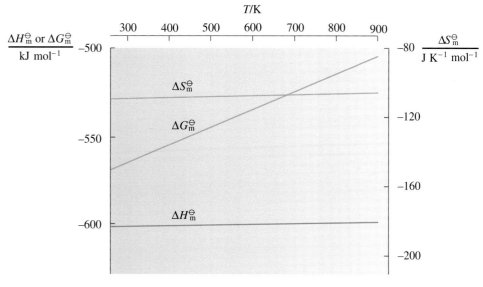

Figure 17.4

Variation with temperature of $\Delta G_m^{\ominus}$, $\Delta H_m^{\ominus}$ and $\Delta S_m^{\ominus}$ for Reaction 2.5. $\Delta G_m^{\ominus}$ and $\Delta H_m^{\ominus}$ values relate to the left-hand axis, and $\Delta S_m^{\ominus}$ values to the right-hand axis.

It follows that in Equation 11.4 we can use the values of $\Delta H_m^{\ominus}$ and $\Delta S_m^{\ominus}$ at 298.15 K, and these values will continue to be appropriate at some higher temperature T. Hence, $\Delta G_m^{\ominus}$ at the higher temperature, $\Delta G_m^{\ominus}(T)$, is given by

$$\Delta G_m^{\ominus}(T) = \Delta H_m^{\ominus}(298.15 \text{ K}) - T\Delta S_m^{\ominus}(298.15 \text{ K}) \tag{17.1}$$

As an illustration, consider the reaction

$$2\text{Ag(s)} + \tfrac{1}{2}\text{O}_2(\text{g}) = \text{Ag}_2\text{O(s)} \tag{17.2}$$

As Table 15.1 shows, at 298.15 K, $\Delta G_m^{\ominus}$ is -11.2 kJ mol^{-1}: this means that Ag$_2$O is thermodynamically stable with respect to silver and oxygen.

● Will $\Delta S_m^{\ominus}$ for this reaction be positive or negative?

● There is a decrease in the number of moles of gas, so $\Delta S_m^{\ominus}$ is negative.

● By referring to Equation 17.1, estimate whether $\Delta G_m^{\ominus}$ for Reaction 17.2 will increase or decrease with increasing temperature.

● Consider the right-hand side of Equation 17.1. The first term, $\Delta H_m^{\ominus}(298.15 \text{ K})$, is a constant: it does not change with temperature. Because $\Delta S_m^{\ominus}(298.15 \text{ K})$ is negative, the second term, $-T\Delta S_m^{\ominus}(298.15 \text{ K})$, is positive. As T increases, the second term will therefore become more positive. We conclude therefore that $\Delta G_m^{\ominus}$ for Reaction 17.2 will become more positive as T increases.

At some higher temperature, $\Delta G_m^{\ominus}$ will change from a negative to a positive quantity. At this temperature, Ag$_2$O will cease to be thermodynamically stable with respect to silver and oxygen, and it will become thermodynamically unstable. If the reaction is fast enough, decomposition will then occur. The changeover temperature, the temperature at which $\Delta G_m^{\ominus}$ becomes zero, is called the **thermodynamic decomposition temperature**.

● Use Equation 17.1 to obtain a formula for this decomposition temperature.

● When $\Delta G_m^{\ominus}(T) = 0$,

$$\Delta H_m^{\ominus}(298.15 \text{ K}) = T\Delta S_m^{\ominus}(298.15 \text{ K})$$

$$T = \frac{\Delta H_m^{\ominus}(298.15 \text{ K})}{\Delta S_m^{\ominus}(298.15 \text{ K})} \tag{17.3}$$

Table 17.1 Thermodynamic data for the silver–oxygen reaction at 298.15 K

	$\Delta G_m^{\ominus}$/kJ mol^{-1}	$\Delta H_m^{\ominus}$/kJ mol^{-1}	$\Delta S_m^{\ominus}$/J K^{-1} mol^{-1}
$2\text{Ag(s)} + \tfrac{1}{2}\text{O}_2(\text{g}) = \text{Ag}_2\text{O(s)}$	-11.2	-31.0	-66.5

Table 17.1 lists data for the silver–oxygen reaction at 298.15 K. Using these in Equation 17.3, we obtain

$$T = \frac{-31.0 \text{ kJ mol}^{-1}}{-66.5 \text{ J K}^{-1} \text{ mol}^{-1}} = \frac{31000 \text{ J mol}^{-1}}{66.5 \text{ J K}^{-1} \text{ mol}^{-1}} = 466 \text{ K}$$

Notice how $\Delta H_m^{\ominus}$, which is quoted in tables in kJ mol^{-1}, must be converted to J mol^{-1} to match the J in the unit of $\Delta S_m^{\ominus}$.

This prediction is confirmed experimentally: when it is heated, silver oxide does indeed decompose at about 466 K. For Reaction 17.2, $\Delta G_m^{\ominus}$ increases with temperature because $\Delta S_m^{\ominus}(298.15\,\text{K})$ is negative. This conclusion is perfectly general: when the temperature is increased, $\Delta G_m^{\ominus}$ for any reaction becomes more positive if $\Delta S_m^{\ominus}$ is negative, and more negative if $\Delta S_m^{\ominus}$ is positive.

Now look again at Table 15.1.

● Will $\Delta G_m^{\ominus}$ for all these reactions become more positive as the temperature increases?

● Yes. For each reaction there is a decrease in the number of moles of gas: $\Delta S_m^{\ominus}$ will be negative in all cases.

In Figure 17.5 we plot $\Delta G_m^{\ominus}$ for selected reactions from Table 15.1 against T. Such a plot is called an **Ellingham diagram**. As we predicted, in each case $\Delta G_m^{\ominus}$ becomes more positive as T increases. Ag_2O appears right at the top of the Figure, and $\Delta G_m^{\ominus}$ does become zero close to our calculated temperature of 466 K.

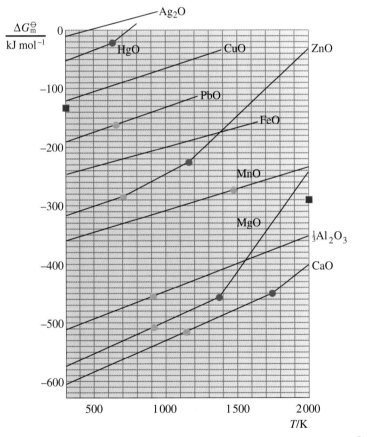

Figure 17.5

An Ellingham diagram showing values of $\Delta G_m^{\ominus}$ for the formation of metal oxides from the metallic element and half a mole of $O_2(g)$ between 298 K and 2 000 K. Blue circles represent the points at which the metals boil, green circles the points at which they melt. The red squares on the vertical axis are for the reaction

$$C(\text{graphite}) + \tfrac{1}{2}O_2(g) = CO(g)$$

Immediately below the plot for Ag_2O is the one for HgO. In this case, $\Delta G_m^{\ominus}$ becomes zero at about 750 K. You can see that the decomposition temperature would be higher than this, were it not for a steepening of the slope above the blue circle at about 630 K. Blue circles in Figure 17.5 represent points at which the metals boil: at 630 K, the mercury in the equation

$$Hg + \tfrac{1}{2}O_2(g) = HgO(s) \tag{17.4}$$

changes from a liquid to a gas.

● Why does this increase the upward slope of the $\Delta G_m^\ominus$ plot?

● Above 630 K, there is a decrease of 1.5 mol of gas in the equation, so $\Delta S_m^\ominus$ becomes more negative. The rate of increase of $\Delta G_m^\ominus$ with T therefore becomes greater. At about 750 K, $\Delta G_m^\ominus$ is zero and HgO decomposes to oxygen gas and mercury vapour.

HgO and Ag_2O undergo thermal decomposition at lowish temperatures because they are near the bottom of Table 15.1. The $\Delta G_m^\ominus$ values for their formation reactions are not very negative, so the increase in $\Delta G_m^\ominus$ with temperature (a consequence of negative $\Delta S_m^\ominus$ values) brings $\Delta G_m^\ominus$ to zero at a relatively low value of T. However, this is not the case for the other oxides in Figure 17.5. These decompose at inconveniently high temperatures, well above 1 000 K. To reduce these oxides to the metal, some method other than thermal decomposition must be used.

17.2 Carbon as a reducing agent

Because of the way our Ellingham diagram is plotted, values of $\Delta G_m^\ominus$ for the oxides at the top of Table 15.1 occur at the bottom of Figure 17.5. This means that if, at some temperature T, the $\Delta G_m^\ominus$ plot for the oxide of metal A lies below that for the oxide of metal B, then metal A is thermodynamically capable of reducing the oxide of metal B at that temperature. Now the lines in Figure 17.5 tend to have similar slopes, except where the boiling of metals causes an abrupt change *. This means that the *relative* strengths of the metals as reducing agents change little with temperature: calcium, for example, is thermodynamically capable of reducing all the other oxides, both at 298 K and 2 000 K. We can attribute the similar slopes to similar negative values of $\Delta S_m^\ominus$: all the formation reactions, when the metal is solid, involve a decrease of half a mole of oxygen gas. When the metal becomes gaseous, there is then a total decrease of 1.5 mol.

Now consider the gas carbon monoxide, whose formation reaction is

$$C(s) + \frac{1}{2}O_2(g) = CO(g) \tag{17.5}$$

● If a plot for this oxide is inserted in Figure 17.5, will its slope resemble that of the other oxides?

● In the formation reaction, there is an *increase* of half a mole of gas, so $\Delta S_m^\ominus$ is *positive* and $\Delta G_m^\ominus$ *decreases* with temperature. The plot for carbon monoxide on the Ellingham diagram will therefore slope downwards from left to right.

You can obtain this plot by joining the two red squares at the extreme left and right of Figure 17.5.

● Join them with a pencilled line. If carbon forms CO(g), how many of the oxides is it thermodynamically capable of reducing at 298.15 K?

● Three. On the left-hand axis, only silver oxide, mercury oxide and copper oxide lie above carbon monoxide.

---

* The melting of metals also causes a change in slope, but as Figure 17.5 shows, the change is very slight.

○ How many of the oxides is carbon thermodynamically capable of reducing at 2 000 K?

○ At 2 000 K, all of the oxide plots except those for CaO and Al_2O_3 lie above the carbon monoxide line. Hence all except these oxides can be reduced by carbon at this temperature.

When carbon is used to reduce oxides, the temperature is usually such that $CO(g)$ is produced rather than $CO_2(g)$. As we have seen, carbon then becomes a more and more powerful reducing agent as the temperature is raised. The ability of carbon to reduce magnesium oxide, provided the reaction temperature is about 2 000 K, is especially striking. However, such temperatures require expensive and sophisticated furnace technology: consequently, carbon reduction is normally used at operating temperatures below about 1 750 K. At this temperature, carbon going to carbon monoxide will still reduce the oxides of lead, zinc and, most important of all, iron.

The deliberate use of carbon to reduce oxygen compounds of copper and iron seems to have occurred first somewhere in eastern Europe or the Middle East. For copper, the date was about 4500 BC; for iron, about 1200 BC. With the possible exception of agriculture, a more important technological step along the road to modern civilization can hardly be imagined (Figure 17.6). It should now be clear to you that this technology is based on a unique property of carbon: it is a highly involatile element which forms volatile oxides. This combination yields the positive value (89.4 J K^{-1} mol^{-1}) of $\Delta S_m^{\ominus}$ for the carbon monoxide formation reaction.

Figure 17.6
A reconstruction of primitive iron production.

17.3 A survey of metal extraction methods

We have now covered the reduction of the oxides of metals in our first and second classes. In the third class are the remaining metals, which include the alkali and alkaline earth metals, and aluminium. These have oxides that either require inconveniently high furnace temperatures for carbon reduction, or are difficult to prepare and handle. Electrolytic methods are widely used, often on fused halides.

It should now be clear that, because of the broad similarities between different series, like those set out in Tables 12.1 and 15.1, and in the answer to Question 15.2,

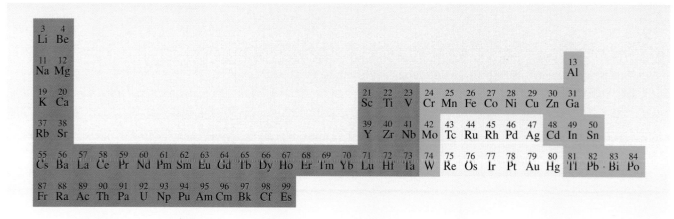

Figure 17.7 An approximate classification of metals by extraction processes.

it is possible to say that, in a thermodynamic sense, some metals such as the alkali and alkaline earth metals are much more difficult to extract from their ores than others. This statement can be made without paying detailed attention to the precise nature of the ore being discussed. The classes 1–3 described at the beginning of Section 17 serve to arrange groups of important metals in order of increasing difficulty of extraction, and relate certain practical features of metal extraction to thermodynamic series such as that given in Table 15.1.

In Figure 17.7, an attempt has been made to extend the classification to all the metals in the Periodic Table; it is only approximate. Broadly speaking, the generalizations made in Figure 17.7 are correct, but some metals require rather specific extraction methods that defy attempts at straightforward generalizations. The shading key to Figure 17.7 is:

Red These metals occur towards the top of series such as those in Tables 12.1 and 15.1. They are usually extracted by the electrolysis of fused halides, or by heating oxides or halides with a strongly reducing metal such as calcium. For example, some alkali and alkaline earth metals are obtained by the electrolysis of fused chlorides. The lanthanide metals and actinide metals, such as uranium, are extracted by heating fluorides with calcium. All these metals usually dissolve quickly in dilute acids, forming positive ions and evolving hydrogen gas.

Green Most of these metals are extracted by carbon reduction of oxides. They nearly all (note the exception of copper) dissolve slowly in dilute hydrochloric acid at room temperature, forming positive ions and evolving hydrogen gas.

Yellow These metals occur towards the bottom of series such as those in Tables 12.1 and 15.1. They occur in the elemental form, or as ores from which the metal can usually be obtained by moderate heating in air. They do not dissolve in dilute hydrochloric acid.

17.4 Summary of Section 17

1   In Equation 11.4, $\Delta H_m^{\ominus}$ and $\Delta G_m^{\ominus}$ do not vary much with temperature, so, to a good approximation, we can assume that

$$\Delta G_m^{\ominus}(T) = \Delta H_m^{\ominus}(298.15\,\mathrm{K}) - T\,\Delta S_m^{\ominus}(298.15\,\mathrm{K}) \qquad (17.1)$$

This equation enables us to predict how $\Delta G_m^{\ominus}$ will vary with temperature.

2   An Ellingham diagram for metal oxides shows how $\Delta G_m^{\ominus}$ for the formation of oxides from half a mole of oxygen and the metallic element varies with T. The plots usually have positive slopes because $\Delta S_m^{\ominus}$ is negative.

3   As carbon monoxide is a gas, $\Delta S_m^{\ominus}$ for this oxide is positive. The $\Delta G_m^{\ominus}$ plot has a negative slope, and so carbon is a very powerful reducing agent at high temperatures.

QUESTION 17.1

Use the thermodynamic data at 298.15 K in the *Data Book* to estimate the thermodynamic decomposition temperature for the breakdown of red HgO into mercury vapour and oxygen:

$$\mathrm{HgO(s)} = \mathrm{Hg(g)} + \tfrac{1}{2}\mathrm{O_2(g)} \qquad (17.6)$$

Compare your value with the one implied by Figure 17.5.

$$T = \frac{\Delta H_m^{\ominus}(298.15\,\mathrm{K})}{\Delta S_m^{\ominus}(298.15\,\mathrm{K})} = 734\,\mathrm{K}$$

QUESTION 17.2

Use Figure 17.5 to find the approximate temperatures at which carbon being oxidized to CO(g) will reduce MnO and ZnO. Write an equation for the reaction in the zinc case, specifying the reactants and products at the reduction temperature.

$$1750\,\mathrm{K} + 1200\,\mathrm{K}$$
$$\mathrm{C(s)} + \mathrm{ZnO(s)} = \mathrm{CO(g)} + \mathrm{Zn(g)}$$

QUESTION 17.3

Tick the appropriate column to show whether the following statements are true or false:

		True	False
(i)	Thermodynamically speaking, copper is harder to extract from its oxide than tin.		✓
(ii)	Carbon going to carbon monoxide becomes a better reducing agent for metal oxides as the temperature is raised.	✓	
(iii)	The lanthanide and actinide elements are found free in nature.		✓
(iv)	Calcium metal should be a good agent for reducing the compounds of other metals, whether those compounds are oxides, fluorides, chlorides, bromides or iodides.	✓	
(v)	The alkali metals are obtained by heating their oxides or chlorides.		✓
(vi)	Electrolysis is often used to extract the metals that form compounds with the most negative values of $\Delta G_f^{\ominus}$.	✓	

THE BORN–HABER CYCLE

18

In Tables 12.1 and 15.1, and in the chloride series in the answer to Question 15.2, metals were arranged in order of their thermodynamic tendency to combine with particular oxidizing agents. In Section 15, we pointed out that these series differed in detail, but also noted that there were marked *overall* similarities between the series, for example the high positions of the alkali metals, and the low positions of copper, silver and gold. What is it about the alkali metals that gives them a high position in each series? We shall now try to answer this, the last of our four introductory questions.

Let us compare sodium and silver chlorides. At 298.15 K

$$Na(s) + \tfrac{1}{2}Cl_2(g) = NaCl(s); \quad \Delta G_m^{\ominus} = -384 \text{ kJ mol}^{-1} \tag{5.2}$$

$$Ag(s) + \tfrac{1}{2}Cl_2(g) = AgCl(s); \quad \Delta G_m^{\ominus} = -110 \text{ kJ mol}^{-1} \tag{18.1}$$

As in the other series, $\Delta G_m^{\ominus}$ is more negative for the sodium than for the silver reaction, and sodium lies above silver. Now the two reactions are analogous reactions in the sense defined in Section 16. The $T\Delta S_m^{\ominus}$ values will therefore be similar, and because

$$\Delta G_m^{\ominus} = \Delta H_m^{\ominus} - T\Delta S_m^{\ominus} \tag{11.4}$$

the difference in the $\Delta G_m^{\ominus}$ values for the two reactions will be almost entirely due to the difference in $\Delta H_m^{\ominus}$. The *Data Book* gives us $\Delta H_m^{\ominus}$ values for the two reactions:

$$Na(s) + \tfrac{1}{2}Cl_2(g) = NaCl(s); \quad \Delta H_m^{\ominus} = -411 \text{ kJ mol}^{-1} \tag{5.2}$$

$$Ag(s) + \tfrac{1}{2}Cl_2(g) = AgCl(s); \quad \Delta H_m^{\ominus} = -127 \text{ kJ mol}^{-1} \tag{18.1}$$

- In what way do the $\Delta G_m^{\ominus}$ and $\Delta H_m^{\ominus}$ values bear out our assertion that the two $T\Delta S_m^{\ominus}$ terms will be similar?

- The difference in the $\Delta G_m^{\ominus}$ values (274 kJ mol^{-1}) is similar to the difference in the $\Delta H_m^{\ominus}$ values (284 kJ mol^{-1}).

It follows that to explain why sodium lies above silver in the chloride series of Question 15.2, our main problem is to explain why $\Delta H_m^{\ominus}$ is more negative for the sodium reaction. We shall try to do this by relating the $\Delta H_m^{\ominus}$ values to the thermodynamic properties of sodium and silver atoms.

You know that one mole of sodium metal and half a mole of chlorine gas react directly to give one mole of sodium chloride via Equation 5.2. However, we shall now imagine an alternative, indirect route at 298.15 K, which takes place in steps of our own choosing, and consider the enthalpy change for each step. These steps are as follows:

1 The metallic bonds in the sodium metal are broken and the metal is converted into gaseous sodium atoms:

$$Na(s) = Na(g) \tag{18.2}$$

The standard enthalpy change for this step is positive and is called the **standard enthalpy of atomization** of sodium, $\Delta H_{atm}^{\ominus}(Na, s)$. It is 107 kJ mol^{-1}.

2 The bonds in our half a mole of Cl_2 molecules are broken, converting the molecules into chlorine atoms:

$$\tfrac{1}{2}Cl_2(g) = Cl(g) \tag{18.3}$$

The standard enthalpy change for this step is positive. When one mole of Cl_2 molecules decompose in this way, the standard enthalpy change is

$$Cl_2(g) = 2Cl(g); \; \Delta H_m^{\ominus} = 244\,kJ\,mol^{-1} \tag{18.4}$$

It is then called the **bond dissociation energy** of the chlorine molecule and given the symbol $D(Cl-Cl)$. For Reaction 18.3, the standard enthalpy change is *half* the bond dissociation energy of the chlorine molecule, $\tfrac{1}{2}D(Cl-Cl)$. This is $122\,kJ\,mol^{-1}$.

3 Now an electron is removed from each gaseous sodium atom:

$$Na(g) = Na^+(g) + e^-(g) \tag{18.5}$$

The standard enthalpy change at 298.15 K for this step is positive ; it is referred to as the **first ionization energy** of sodium, $I_1(Na)$, and its value is $496\,kJ\,mol^{-1}$.

4 Now the electrons from the sodium atoms are added to the gaseous chlorine atoms to form gaseous chloride ions:

$$Cl(g) + e^-(g) = Cl^-(g) \tag{18.6}$$

Here you have to be careful. In Reaction 18.6, heat is evolved, and traditionally, the energies of it, and reactions like it, are defined in terms of an *electron affinity*, in this case the electron affinity of chlorine, $E(Cl)$, which is the amount of heat evolved when Reaction 18.6 takes place. Here too therefore, the **electron affinity** of an element is defined as the heat evolved when a gaseous atom of the element accepts an electron at 298.15 K. Now we require the *enthalpy change* of Reaction 18.6, which, from Equation 6.1 is the heat absorbed. This is equal to minus the heat evolved, and as the electron affinity is the heat evolved, $\Delta H_m^{\ominus} = -E(Cl)$.

● The electron affinity of chlorine is $349\,kJ\,mol^{-1}$. What is the value of $\Delta H_m^{\ominus}$ for Reaction 18.6?

● $\Delta H_m^{\ominus} = -E(Cl) = -349\,kJ\,mol^{-1}$

5 The overall effect of steps 1–4 has been to convert sodium metal and chlorine gas into the gaseous ions, $Na^+(g)$ and $Cl^-(g)$. This can be seen by adding Equations 18.2, 18.3, 18.5 and 18.6:

$$Na(s) + \tfrac{1}{2}Cl_2(g) = Na^+(g) + Cl^-(g) \tag{18.7}$$

So the final step on the way to NaCl(s) is the one in which these gaseous ions come together to form solid crystalline sodium chloride:

$$Na^+(g) + Cl^-(g) = NaCl(s) \tag{18.8}$$

The standard enthalpy change of this reaction at 298.15 K is called the **lattice energy** of sodium chloride and given the symbol $L(NaCl, s)$. Equation 18.8 is a process of bond formation prompted by the attractive forces between oppositely charged ions, so it is one in which heat is evolved. Thus, $L(NaCl, s)$ is negative, but how negative is it?

We can find out by representing what we have done in a *thermodynamic cycle*. This particular one is known as the **Born–Haber cycle** (Figure 18.1).

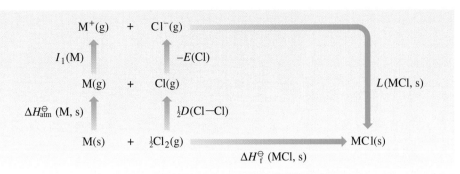

Figure 18.1
Born–Haber cycle for a solid univalent metal chloride, MCl.

A solid chloride, MCl, is formed from the metal and chlorine either by the single equation at the base of the cycle with an arrow in the anticlockwise direction or by the clockwise sequence of steps 1–5 that have been described above. Where M is sodium or silver, all the terms in the cycle have been individually determined by experimental methods except for the lattice energy, L. The values are shown in Table 18.1.

Table 18.1 Terms in the Born–Haber cycles for sodium and silver chlorides

Term	NaCl	AgCl
$\Delta H^{\ominus}_{atm}(M, s)/kJ\, mol^{-1}$	107	285
$I_1(M)/kJ\, mol^{-1}$	496	731
$\frac{1}{2}D(Cl-Cl)/kJ\, mol^{-1}$	122	122
$-E(Cl)/kJ\, mol^{-1}$	−349	−349
$L(MCl, s)/kJ\, mol^{-1}$	−787	−916
$\Delta H^{\ominus}_{f}(MCl, s)/kJ\, mol^{-1}$	−411	−127

🔵 Write down an equation that relates $\Delta H^{\ominus}_{f}(MCl, s)$ to the other terms in the cycle.

🔵 The cycle describes the formation of MCl from the metal and chlorine, either by the equation at the base of the cycle with an arrow in the anticlockwise direction, or by the five-step clockwise route described in this Section. By Hess's law, the enthalpies of the two routes must be equal. So

$$\Delta H^{\ominus}_{f}(MCl, s) = \Delta H^{\ominus}_{atm}(M, s) + I_1(M) + \frac{1}{2}D(Cl-Cl) - E(Cl) + L(MCl, s) \quad (18.9)$$

🔵 Now use the data in Table 18.1 to calculate the lattice energy of sodium chloride.

🔵 From Equation 18.9, the top five numbers in Table 18.1 must add up to the bottom number. Thus, for NaCl:

$$(107 + 496 + 122 - 349)\, kJ\, mol^{-1} + L(NaCl, s) = -411\, kJ\, mol^{-1}$$

$$L(NaCl, s) = -787\, kJ\, mol^{-1}$$

If this calculation is repeated with the data for silver, one finds that $L(AgCl, s) = -916\, kJ\, mol^{-1}$. Pencil those two lattice energies into the blank spaces in Table 18.1.

18.1 Comparing sodium and silver halides

You should now have a complete Table 18.1, which shows how $\Delta H_f^{\ominus}$ values for NaCl(s) and AgCl(s) can each be obtained by adding up five quantities. Of these five quantities, two, $\frac{1}{2}D(Cl-Cl)$ and $-E(Cl)$, are identical for both NaCl and AgCl. This leaves $\Delta H_{atm}^{\ominus}(M, s)$, $I_1(M)$ and $L(MCl, s)$.

● Which of these three quantities tend to make $\Delta H_f^{\ominus}$ more negative for NaCl than for AgCl?

● The standard enthalpy of atomization and the first ionization energy of the metal: $\Delta H_{atm}^{\ominus}(Na, s)$ is 178 kJ mol^{-1} less than $\Delta H_{atm}^{\ominus}(Ag, s)$; and $I_1(Na)$ is 235 kJ mol^{-1} less than $I_1(Ag)$. Equation 18.9 shows that both these terms tend to make $\Delta H_f^{\ominus}(NaCl, s)$ more negative than $\Delta H_f^{\ominus}(AgCl, s)$. The other term, the lattice energy, tends to produce the *opposite* effect, because it is 129 kJ mol^{-1} more negative for AgCl than for NaCl.

Thus, if we parcel up the enthalpies of formation of the two chlorides in the way that we have done in this Section, we find that it is mainly the larger value of the sum $(\Delta H_{atm}^{\ominus}(M, s) + I_1)$ which causes silver chloride to lie below sodium chloride in a series based on the values of $\Delta H_m^{\ominus}$ for Reactions 5.2 and 18.1. As the two values of $T\Delta S_m^{\ominus}$ are so similar, silver chloride also lies below sodium chloride in a series based on the values of $\Delta G_m^{\ominus}$.

Now suppose that we changed the reactant that combined with the metal M in Figure 18.1 from chlorine to, say, fluorine, and redrew the cycle.

● What thermodynamic quantities in the cycle would remain unchanged?

● The terms that are properties *of the metal alone*, $\Delta H_{atm}^{\ominus}(M, s)$ and $I_1(M)$. If we consider a change from NaCl to NaF, $\frac{1}{2}D(Cl-Cl)$ would change to $\frac{1}{2}D(F-F)$, $E(Cl)$ would change to $E(F)$, and $L(NaCl, s)$ would change to $L(NaF, s)$. However, $\Delta H_{atm}^{\ominus}(Na, s)$ and $I_1(Na)$ would remain the same.

It follows that, in a reconstituted Table 18.1 that compared NaF with AgF, the top two rows, which give $\Delta H_{atm}^{\ominus}(M, s)$ and $I_1(M)$ values, would remain the same. These values would now tend to make $\Delta H_f^{\ominus}(NaF, s)$ more negative than $\Delta H_f^{\ominus}(AgF, s)$. If we generalize this observation, we see that the $\Delta H_{atm}^{\ominus}(M, s)$ and $I_1(M)$ terms exert such an influence on the relative values of the standard enthalpies of formation, that there will be a strong tendency for sodium to be above silver whether the metals react with chlorine, fluorine, or any other standard reactant. Basically, then, the alkali and alkaline earth metals lie above noble metals such as copper, mercury, silver and gold in Tables 12.1 and 15.1, because they have lower atomization enthalpies and lower ionization energies. As we switch from one Table to another, variations in lattice energies produce important changes in detail, but the noble metals always lie near the bottom, and the alkali and alkaline earth metals are always near the top.

In conclusion then, metals with relatively low enthalpies of atomization and relatively low ionization energies are those that are hardest to extract from their compounds. They are also the metals that most readily reduce other metallic compounds or ions.

18.2 Summary of Section 18

1 The Born–Haber cycle enables us to calculate the lattice energies of halides from experimental data such as the standard enthalpies of atomization and ionization energies of metals, the dissociation energies and electron affinities of halogens and the enthalpies of formation of the halides.

2 Analysis via the Born–Haber cycle suggests that metals with relatively low enthalpies of atomization and ionization energies are hardest to extract from their compounds.

QUESTION 18.1

Figure 18.2 shows the Born–Haber cycle for lithium fluoride. The quantities a–f represent enthalpy changes for reactions occurring in the direction of the arrows. From the quantities listed below, choose those that should be assigned to the six steps a–f in the cycle:

(i)   the dissociation enthalpy of fluorine, $D(F{-}F)$;

(ii)   one-half of the dissociation enthalpy of fluorine, $\frac{1}{2}D(F{-}F)$;

(iii) the electron affinity of fluorine, $E(F)$;

(iv)   minus the electron affinity of fluorine, $-E(F)$;

(v) $-\Delta H_f^{\ominus}(\text{LiF, s})$;

(vi) $\Delta H_f^{\ominus}(\text{LiF, s})$;

(vii) the first ionization energy of the lithium atom, $I_1(\text{Li})$;

(viii) minus the lattice energy of lithium fluoride, $-L(\text{LiF, s})$;

(ix)   the lattice energy of lithium fluoride, $L(\text{LiF, s})$;

(x)   the standard enthalpy of atomization of lithium metal, $\Delta H_{\text{atm}}^{\ominus}(\text{Li, s})$.

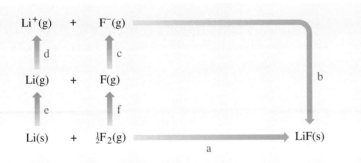

Figure 18.2 Born–Haber cycle for lithium fluoride, LiF.

QUESTION 18.2

Use the following data to calculate the lattice energy of lithium fluoride:

$\Delta H_{\text{atm}}^{\ominus}(\text{Li, s}) = 159\,\text{kJ mol}^{-1}$;

$I_1(\text{Li}) = 520\,\text{kJ mol}^{-1}$;

$D(F{-}F) = 158\,\text{kJ mol}^{-1}$;

$E(F) = 328\,\text{kJ mol}^{-1}$;

$\Delta H_f^{\ominus}(\text{LiF, s}) = -616\,\text{kJ mol}^{-1}$.

QUESTION 18.3

The value of $\Delta H_f^{\ominus}$ (CaCl$_2$, s) is −796 kJ mol^{-1}. For calcium, the standard   −2255
enthalpy of atomization of the metal is 178 kJ mol^{-1}, and the first and second
ionization energies are 590 and 1 145 kJ mol^{-1}, respectively. Use these data, and
data from Table 18.1, to calculate a value for the lattice energy of solid CaCl$_2$.
Compare your value with that for NaCl in Section 18, and comment on the
difference.

QUESTION 18.4

Look at Figure 17.7. Broadly speaking, would you expect those metals with the
highest ionization energies to be to the left or the right of Figure 17.7?   RISHT

INTRODUCTION TO THE REMAINING SECTIONS

19

We have now completed our introduction to thermodynamics and, at the same time, responded to the problem of the activity series of metals. Metals such as sodium and calcium are powerful reducing agents, and hard to extract from their ores. They have this status because of their relatively low enthalpies of atomization and ionization energies. The chief purpose of the rest of this Book is to begin a study of the chemistry of the typical elements. So between Section 23 and the end of the Book, you will find an account of the chemistry of the elements of Group I and Group II — he alkali and alkaline earth metals. But before we can embark on the descriptive chemistry, there are three thermodynamic quantities that you need to know more about. You have already been introduced to two of them — the lattice energy and the ionization energies of elements, but we need to add some more details. The third quantity is called the redox potential, and it provides an 'alternative currency' for the compiling of Table 12.1. Sections 20–22 are devoted to these three thermodynamic quantities.

THE LATTICE ENERGY

In Section 18 and Question 18.3, we used the Born–Haber cycle to calculate the lattice energies of sodium and calcium chlorides, $L(\text{NaCl, s})$ and $L(\text{CaCl}_2, \text{s})$. They are equal to the standard enthalpy changes of the following reactions:

$$\text{Na}^+(\text{g}) + \text{Cl}^-(\text{g}) = \text{NaCl(s)}; \ \Delta H_m^{\ominus} = -787 \text{ kJ mol}^{-1} \qquad (20.1)$$

$$\text{Ca}^{2+}(\text{g}) + 2\text{Cl}^-(\text{g}) = \text{CaCl}_2(\text{s}); \ \Delta H_m^{\ominus} = -2\,255 \text{ kJ mol}^{-1} \qquad (20.2)$$

In these reactions, gaseous ions condense to form a solid. No one has yet succeeded in determining the energies of such changes by direct measurements on reactions such as 20.1 and 20.2. The only way of obtaining experimental values of lattice energies is through the Born–Haber cycle. In this Section we shall seek reasons why these important quantities have the values that they do.

Sodium and calcium chlorides are usually treated as ionic compounds: they consist of charged ions. To pull oppositely charged ions such as Na^+ and Cl^- apart against the attractive force of the ions requires work (energy). So the separation of ions is an endothermic process and ΔH is positive.

- How does this explain the negative lattice energies of NaCl and CaCl_2 in Equations 20.1 and 20.2?

- When the gaseous, oppositely charged ions are *brought together* to form a solid, energy is released and the process is exothermic.

These arguments are applications of an **ionic model** (Figure 20.1), a model which assumes that certain compounds are composed of ions. They are also only qualitative. But in the next Section we shall express them quantitatively.

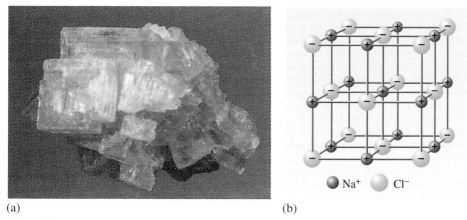

(a)   (b)

Figure 20.1 The ionic model represents a crystal (a) of halite (sodium chloride) as an assembly of oppositely charged ions, like that used in (b) to represent the structure's unit cell.

20.1 The energy of interaction between two ions

A quantitative application of the ionic model is discussed in *The Third Dimension*[2]. The underlying assumption was that ions could be represented by hard charged spheres. This assumption allows a set of ionic radii to be developed, and then used to predict likely crystal structures from radius ratios. We shall use the same notion of hard charged spheres here.

Suppose we have two such ions (e.g. Na^+ and Cl^-) with equal and opposite single charges $+e$ and $-e$; the centres of the ions are separated by a distance r (Figure 20.2). Because of the opposite charges, there is a Coulomb force of attraction between the ions. It then turns out that the electrostatic *energy*, E_C, arising from this Coulomb attraction is given by:

$$E_C = -\frac{e^2}{4\pi\varepsilon_0 r} \tag{20.3}$$

Here e is the electronic charge, 1.6×10^{-19} C, and ε_0 is a known physical constant called the *permittivity of a vacuum*. Thus, on the top of the fraction in Equation 20.3 we have the product of the equal and opposite charges, $+e$ and $-e$; on the bottom, we have a constant $(4\pi\varepsilon_0)$ multiplied by the internuclear distance, r.

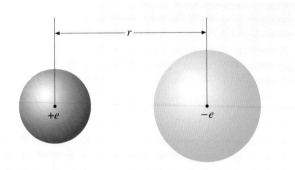

Figure 20.2 Two hard oppositely charged spheres, whose centres are separated by a distance r.

Now imagine that the two charged spheres are separated by an enormous distance — so large that it may be regarded as infinite.

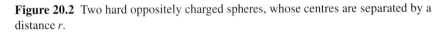 What then is the value of E_C?

Zero; the fraction in Equation 20.3 consists of a quantity, $-e^2$, divided by $4\pi\varepsilon_0 r$, where r is infinitely large. It will therefore be minute — effectively zero.

Now suppose the two ions, initially separated by this huge distance, move progressively closer together. The separation r gets smaller, so the quantity $(e^2/4\pi\varepsilon_0 r)$ will increase from zero. But in Equation 20.3, this quantity has a negative sign in front of it, so E_C, from being zero, will gradually become more negative. The steady decrease in E_C occurs naturally because the equal and opposite charges are pulled together by an attractive force, so they tend to approach each other.

⬤ When does this natural approach of the two ions stop?

⬤ We are modelling our ions as hard spheres. It will stop when they make contact with each other (Figure 20.3).

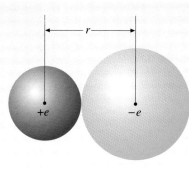

Figure 20.3
Two oppositely charged ions, represented as hard spheres, in contact with each other. The internuclear distance r is now the sum of their radii.

You can think of the spheres as balls made of hard rubber. When they make contact, the steady decrease in potential energy caused by the decrease in E_C stops. Any further decrease in the internuclear distance r must be brought about by pressing the spheres together against the repulsive force of the hard rubber. This requires us to do work on them, so there is now a steep increase in the energy of interaction due to a new term, the repulsive energy E_R. The total energy for the two ions is therefore

$$E = E_C + E_R \qquad (20.4)$$

The combined effect of these two terms is shown in Figure 20.4. To the right, the Coulombic energy, E_C, is dominant and as the internuclear distance, r, diminishes, E becomes more negative. To the left, after the ions have come into contact, E_R is dominant and E increases. In between, there is an equilibrium internuclear distance r_0, where the total energy of interaction, E, is a minimum, E_{min}. This is the most stable situation, and the internuclear distance observed at this energy minimum corresponds to that found in crystals.

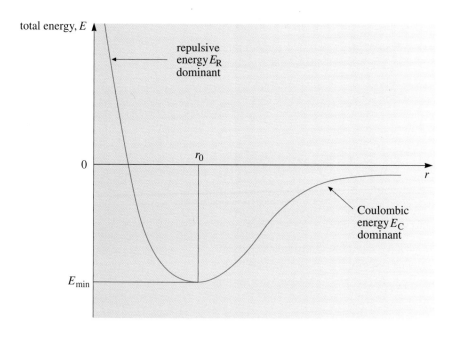

Figure 20.4
How the total interaction energy, E, of two oppositely charged ions varies with internuclear distance.

BOX 20.1 Max Born (1882–1970)

Although Max Born (Figure 20.5) contributed to the theory of ionic crystals through the Born–Haber cycle and Born exponent, his most fundamental work was on statistical interpretations of the quantum theory. The idea that the distribution of electrons in atoms is a distribution of probabilities was essentially his. This work was done mainly at Göttingen in 1921–1933, when he worked with Heisenberg, Fermi, Hund, J. R. Oppenheimer, Pauli, Teller, von Neumann and Wigner. In 1933, Nazi pressure on Jews led him to leave Germany, and from 1936–1953, he was Tait Professor at Edinburgh. His political experiences left him deeply pessimistic. He was an enthusiastic supporter of Germany's cause in the First World War, but in the Cold War period of the 1950s he was a leading campaigner against the development of nuclear weapons.

Figure 20.5
Max Born (1882–1970).

In Equation 20.4, we have the two terms E_C and E_R on the right. We already have a quantitative expression for E_C in Equation 20.3. To go further, we need one for E_R also. Max Born (Figure 20.5) suggested that one possible answer has the form

$$E_R = \frac{B}{r^n} \qquad (20.5)$$

where B is a constant, and n (which is known as the Born exponent) is a number lying between 5 and 12. Thus, E_R is proportional to $1/r^n$, where n is much greater than one. This is why E_R increases so steeply at short internuclear distances at the left of the minimum in Figure 20.4.

We can now substitute Equations 20.3 and 20.5 into Equation 20.4:

$$E = -\frac{e^2}{4\pi\varepsilon_0 r} + \frac{B}{r^n} \qquad (20.6)$$

It is this equation which gives us the curve shown in Figure 20.4. But we are interested in the energy of the minimum, E_{min}. This corresponds to the lattice energy, in that it gives us the energy of the ion pair at the equilibrium internuclear distance r_0, relative to the zero energy they possess as fully separated ions in the gas phase. There are well-established mathematical techniques for finding E_{min}, and at the same time eliminating the constant B. The result is

$$E_{min} = -\frac{e^2}{4\pi\varepsilon_0 r_0}\left(1 - \frac{1}{n}\right) \qquad (20.7)$$

20.2 The lattice energy of an ionic crystal

So far, so good: we have an expression (Equation 20.7) for the energy of interaction of a pair of singly charged ions. But an ionic crystal is more complicated in three important respects:

1   It contains more than two ions! Indeed, we have calculated lattice energies in kJ mol^{-1}. The Avogadro constant, N_A, tells us the number of formula units in one mole. It has the value 6.022×10^{23} mol^{-1}. So when calculating the lattice energy of sodium chloride, we are thinking of crystals containing 6.022×10^{23} Na$^+$ ions and 6.022×10^{23} Cl$^-$ ions.

2   Our ion pair contained singly charged ions, but in general our ionic crystal may contain ions with other integral charges such as $+3e$ or $-2e$.

3   Different ionic crystals have different unit cells, and therefore different arrangements of ions in space. We must expect the expression for the lattice energy to change from one type of structure to another (Figure 20.6).

Allowing for these three complications is mathematically quite difficult, but the result is surprisingly simple. To obtain the lattice energy of an ionic crystal, all that is necessary is to multiply Equation 20.7 by a quantity $N_A M Z_+ Z_-$. Hence

$$L = -\frac{N_A M Z_+ Z_- e^2}{4\pi\varepsilon_0 r_0}\left(1 - \frac{1}{n}\right) \tag{20.8}$$

This is called the *Born–Landé* equation. We have already explained the meaning of the symbols $4\pi\varepsilon_0$, r_0, e and n. N_A is the Avogadro constant, equal to 6.022×10^{23} mol^{-1}; this is the allowance for complication 1 above. Z_+ is the number of positive charges $+e$ on the positive ions in the crystal, and Z_- is the number of negative charges $-e$ on the negative ions; together they allow for complication 2. Finally, M is a number that is characteristic of the particular crystal structure under consideration. It is called the **Madelung constant** and allows for complication 3. Some values of M are shown for various crystal structures in Table 20.1. We shall have more to say about them later.

Figure 20.6
A crystal of fluorite, CaF_2, may contain some 10^{24} ions. The ions have different charges ($+2e$ and $-e$), and a specific arrangement in space. All this must be allowed for when transforming Equation 20.7 into an expression for the lattice energy of the crystal.

Table 20.1 Some crystal data relevant to the calculation of lattice energies

Structure	Compound	Madelung constant, M	Ions in formula unit, ν	M/ν
caesium chloride	CsCl	1.763	2	0.882
sodium chloride	NaCl	1.748	2	0.874
fluorite	CaF_2	2.519	3	0.840
zinc blende	ZnS	1.638	2	0.819
wurtzite	ZnS	1.641	2	0.821
corundum	Al_2O_3	4.172	5	0.835
rutile	TiO_2	2.408	3	0.803

Equation 20.8 can be simplified by substituting numbers or physical constants for N_A, e^2 and $4\pi\varepsilon_0$. The result is

$$L/\text{kJ mol}^{-1} = -\frac{1.389 \times 10^5 M Z_+ Z_-}{r_0}\left(1 - \frac{1}{n}\right)\text{pm} \tag{20.9}$$

Note the units kJ mol^{-1} and pm that appear on the left and right, respectively, of the equation. They tell us that if we insert values of r_0 in picometres (pm), we obtain lattice energies, L, in kJ mol^{-1}.

20.3 Applying the Born–Landé equation

We shall now use Equation 20.9 to calculate the lattice energies of the compounds LiF, NaCl and CaCl$_2$. We shall then compare our results with the experimental values obtained by using the Born–Haber cycle. These experimental values were calculated in Section 18, and in Questions 18.2 and 18.3. They appear in the last column of Table 20.2, along with the quantities that we shall need to substitute into Equation 20.9.

Table 20.2 Calculating the lattice energies of NaCl, NaF and CaCl$_2$ from the Born–Landé equation

Compound	Structure	M	Z_+	Z_-	n	r_0/pm	L/kJ mol^{-1} Born–Landé	Born–Haber cycle
NaCl	NaCl	1.748	1	1	8	281	−756	−787
LiF	NaCl	1.748	1	1	6	201	−1 007	−1 046
CaCl$_2$	rutile	2.408	2	1	9	274	−2 170	−2 255

As the Table shows, NaCl and LiF both have the sodium chloride structure, and CaCl$_2$ has the rutile structure. We can therefore enter values of the Madelung constant M from Table 20.2. Remember that M multiplies up the interaction of the simple ion pair to allow for the fact that we are dealing with the interactions of a collection of ions arranged in a crystal. It is larger for the rutile than for the sodium chloride structure. This is mainly because a mole of TiO$_2$ or CaCl$_2$ crystals contains 50% more ions than a mole of NaCl crystals.

Next in Table 20.2 come the numbers of positive or negative charges on the ions. *Notice that these numbers are all positive.* There are, for example, two positive charges on each Ca^{2+} ion in CaCl$_2$ and one negative charge on each Cl$^-$ ion.

Next come the values of n, the Born exponent. They can be obtained from experiments that study the effect of pressure on the volumes of crystals. However, in most cases we shall be dealing with ions with noble gas configurations. Approximate values of n can then be obtained from the numbers given by Pauling, which appear in Table 20.3. A constant is given for each noble gas configuration. To obtain a value of n for an ionic crystal, one takes the average for the noble gas configurations of the ions in the formula unit. Thus, in NaCl, the Na$^+$ ion has the configuration of neon ($n = 7$), and the Cl$^-$ ion that of argon ($n = 9$), so the average is $n = 8$. This is the value of n for NaCl which appears in Table 20.2.

⬤ What are the values of n for LiF and CaCl$_2$?

⬤ 6 and 9, respectively; in LiF, the Li$^+$ and F$^-$ ions have the helium ($n = 5$) and neon ($n = 7$) configurations so the average is 6. In CaCl$_2$, there are *three* ions to average over, but all of them have the argon configuration so $n = 9$. *Enter these values of 6 and 9 in Table 20.3.*

Table 20.3
Constants used to calculate the Born exponent, n, for crystals containing ions with noble gas configurations

Noble gas configuration	Constant
[He]	5
[Ne]	7
[Ar]	9
[Kr]	10
[Xe]	12

Finally, Table 20.2 contains the values of r_0, the equilibrium internuclear distance that can be obtained by X-ray diffraction. We now have all the data that we need to calculate L from Equation 20.9. For NaCl,

$$L(\text{NaCl, s}) = -\frac{1.389 \times 10^5 \times 1.748 \times 1 \times 1}{281}(1 - \frac{1}{8})\,\text{kJ mol}^{-1}$$
$$= -756\,\text{kJ mol}^{-1}$$

Now do Question 20.1.

QUESTION 20.1

Use Table 20.2, and the data that you have entered in it, to calculate the lattice energies of LiF and $CaCl_2$. Check your results against those given in this Table.

The last two columns of Table 20.2 suggest that Equation 20.9 works pretty well. The values are less negative than the experimental values obtained from the Born–Haber cycle, but only by about 4% or less. Our simple combination of just two terms, E_C and E_R, has 'captured' some 96% of the lattice energy. There are also possible reasons why the calculated values are not negative enough. We have, for example, not allowed for the London (or van der Waals) forces between the ions. Inclusion of these would give more negative values of L and improve the agreement. It looks, therefore, as if our hard-sphere ionic model might well provide us with an understanding of the energies of ionic compounds and their reactions. Let us take it a little further.

20.4 The Kapustinskii equation

Look again at Equation 20.9:

$$L/\text{kJ mol}^{-1} = -\frac{1.389 \times 10^5 MZ_+Z_-}{r_0}(1 - \frac{1}{n})\,\text{pm} \qquad (20.9)$$

To apply it quantitatively to a particular compound, we need the Madelung constant M, which varies from one crystal structure to another. But there are occasions in chemistry when we do not need to be as quantitative as this. It is then useful to have a simpler equation that merely gives us an idea of how lattice energies change from one ionic compound to another, regardless of any change in crystal structure.

Table 20.1 sets us on the road to this simpler equation. Column 3 contains the Madelung constants of various crystal structures, and column 4, the numbers of ions, v, in each formula unit. Thus, in Al_2O_3, there are two Al^{3+} ions and three O^{2-} ions, so $v = 5$. The values of M vary substantially, but in column 5, they have been divided by v.

⬤ What does division by v do to the variation in M?

⬤ It largely disappears. The values of M/v span the small range 0.80–0.88.

This was first noticed by the Russian chemist, A. F. Kapustinskii. What it suggests is that it will not be too bad an approximation to replace M in Equation 20.9 by a number in the range 0.80–0.88, multiplied by v. As Equation 20.9 tends, if anything, to give values that are not quite negative enough, it makes sense to choose a number in the upper part of the range. We therefore choose the NaCl value of 0.874 and replace M by $0.874v$.

Furthermore, you know of a set of ionic radii that can be used to predict internuclear distances in ionic compounds. So we can replace r_0 by $(r_+ + r_-)$, where r_+ and r_- are the radii of the positive and negative ions. The result of these two changes is:

$$L/\text{kJ mol}^{-1} = -\frac{1.214 \times 10^5 v Z_+ Z_-}{r_+ + r_-}\left(1 - \frac{1}{n}\right)\text{pm} \qquad (20.10)$$

One further step is needed. The term $(1 - 1/n)$ in Equation 20.10 consists of two parts. The second part, $1/n$, is only a small fraction of the first, reflecting the fact that the repulsive energy E_R is only a small fraction of the Coulombic energy, E_C. We can therefore approximate this $1/n$ term without much reduction in accuracy. As n varies between 5 and 12 (Table 20.3), Kapustinskii used the value 9. The sum total of these changes gives us the **Kapustinskii equation**:

$$L = -\frac{W v Z_+ Z_-}{r_+ + r_-} \qquad (20.11)$$

where $W = 1.079 \times 10^5\,\text{kJ mol}^{-1}\,\text{pm}$.

For NaCl, LiF and $CaCl_2$, using the ionic radii given in the *Data Book* (on the CD-ROM), this equation gives lattice energies of −763, −1 033 and −2 304 kJ mol⁻¹, respectively. If you look at Table 20.2, you will see that its performance compares very favourably with the more rigorous Born–Landé equation, although when a wider range of compounds is studied, it does not do so well. It will, however, prove very useful in qualitative arguments. For example, the equation shows clearly that $L(\text{LiF, s})$ is more negative than $L(\text{NaCl, s})$ because the ions in LiF are smaller than those in NaCl. Their centres are therefore closer together in the crystal.

20.5 A more general Born–Haber cycle

When we draw the Born–Haber cycle for NaCl (Figure 20.7), it includes the step

$$\tfrac{1}{2}Cl_2(g) = Cl(g); \quad \Delta H_m^{\ominus} = 122\,\text{kJ mol}^{-1} \qquad (18.3)$$

As $Cl_2(g)$ is the standard state of elemental chlorine, and gaseous chlorine atoms are formed in Equation 18.3, we could write the standard enthalpy change of this step as $\Delta H_f^{\ominus}(\text{Cl, g})$. If you look in the *Data Book*, you will see that $\Delta H_f^{\ominus}(\text{Cl, g})$ is indeed 122 kJ mol⁻¹. However, the break up of the molecules in the *gas phase* can be described in terms of bond dissociation energies. As the Cl_2 molecules are gaseous,

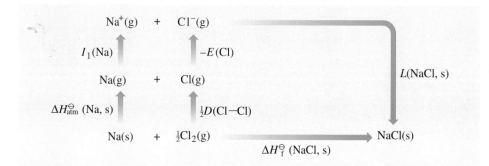

Figure 20.7
The Born–Haber cycle for NaCl.

$\Delta H_f^{\ominus}$ (Cl, g) is also equal to $\frac{1}{2}D$(Cl—Cl), half the bond dissociation energy of Cl_2. It is usual to write the step in this more specific form, as in Figure 20.7.

When we move to NaBr and NaI, this equality between $\Delta H_f^{\ominus}$(X, g) and $\frac{1}{2}D$(X—X) no longer exists. At room temperature, bromine is a liquid and iodine a solid: the standard state of these halogens are Br_2(l) and I_2(s), respectively. The Born–Haber cycle for NaI takes the form shown in Figure 20.8. Again, there is a step corresponding to Equation 18.3:

$$\frac{1}{2}I_2(s) \longrightarrow I(g); \ \Delta H_m^{\ominus} = 107 \, \text{kJ mol}^{-1} \tag{20.12}$$

Again also, $\Delta H_m^{\ominus}$ is equal to $\Delta H_f^{\ominus}$(I, g), as you can verify in the *Data Book*. However, this time it is not equal to $\frac{1}{2}D$(I—I). This is because the diatomic molecule on the left is in the solid rather than the gaseous phase.

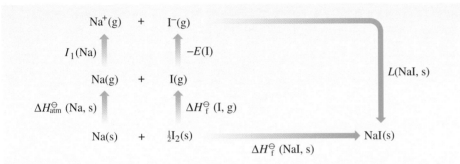

Figure 20.8 The Born–Haber cycle for NaI.

The precise relationship between $\Delta H_f^{\ominus}$ (I, g) and $\frac{1}{2}D$(I—I) is revealed by Figure 20.9.

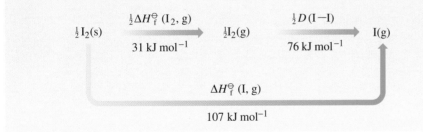
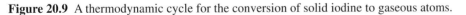

Figure 20.9 A thermodynamic cycle for the conversion of solid iodine to gaseous atoms.

Along the top, Reaction 20.12 takes place in two steps. Firstly, the half-mole of solid iodine is vapourized into gaseous I_2 molecules; the standard enthalpy change is $\frac{1}{2}\Delta H_f^{\ominus}$($I_2$, g). From the *Data Book* this is 31 kJ mol^{-1}. Then the gaseous I_2 molecules are taken apart into gaseous atoms. It is this step whose standard enthalpy change is $\frac{1}{2}D$(I—I), with a value of 76 kJ mol^{-1}. The total energy change from left to right is (31 + 76) or 107 kJ mol^{-1}, and it is this which is equal to $\Delta H_f^{\ominus}$(I, g).

When doing Born–Haber cycle calculations in general, it is best to use a cycle of the type shown in Figure 20.8 with $\Delta H_f^{\ominus}$(X, g) values from the *Data Book*. But you need to be aware of the possible use of bond dissociation energies because — notably for fluorides and chlorides — the data will sometimes be given to you in this form.

20.6 Summary of Section 20

1 A hard-sphere ionic model leads to an expression for the lattice energy of an ionic crystal. This is the Born–Landé equation:

$$L/\text{kJ mol}^{-1} = -\frac{1.389 \times 10^5 \, MZ_+Z_-}{r_0}\left(1-\frac{1}{n}\right)\text{pm}$$

Here, M is a constant called the Madelung constant, which is characteristic of the particular crystal structure. Z_+ and Z_- are the numbers of charges on the positive and negative ions, n is a number known as the Born exponent, and r_0 is the internuclear distance between the ions.

2 The Born–Landé equation gives L values that agree quite closely with the experimental ones obtained from the Born–Haber cycle.

3 The Born–Landé equation can be further simplified into the Kapustinskii equation. This ignores differences in crystal structure, but nevertheless gives a useful idea of how lattice energies vary with the sizes and charges of the ions in ionic compounds. It takes the form

$$L = -\frac{WvZ_+Z_-}{r_+ + r_-}$$

Here, W is a constant equal to $1.079 \times 10^5 \, \text{kJ mol}^{-1}\,\text{pm}$, v is the number of ions in the formula unit, and r_+ and r_- are the radii of the positive and negative ions.

QUESTION 20.2

Draw a Born–Haber cycle for KBr, and use it to write down an equation for the lattice energy of this compound in terms of ionization energy, electron affinity, etc. Substitute data from the *Data Book* into the equation, and thus calculate $L(\text{KBr, s})$.

QUESTION 20.3

KBr has the sodium chloride structure, with an internuclear distance of 330 pm. Use the Born–Landé equation to calculate its lattice energy. How well does your value agree with the experimental value that you obtained in Question 20.2?

QUESTION 20.4

Estimate the Born exponent, n, for CaF_2, which has the fluorite structure, with an internuclear distance of 237 pm. Calculate $L(\text{CaF}_2,\text{s})$ from the Born–Landé equation. How well does your value compare with the experimental figure of $-2\,635 \, \text{kJ mol}^{-1}$ obtained from the Born–Haber cycle?

QUESTION 20.5

The values for the ionic radii of Ca^{2+} and F^- given in the *Data Book* are 114 pm and 119 pm, respectively. Use the Kapustinskii equation to calculate $L(\text{CaF}_2,\text{s})$. Is the agreement with the experimental value better or worse than that obtained with the Born–Landé equation?

ELECTROCHEMICAL CELLS AND REDOX POTENTIALS

21

(Pg 95)

In Table 12.1, we graded metals in terms of their strengths as reducing agents in aqueous solution. Suppose a metal M forms an aqueous ion $M^{n+}(aq)$.

⬤ What then is the reaction whose $\Delta G_m^{\ominus}$ value appears in Table 12.1?

⬤ $\frac{1}{n}M(s) + H^+(aq) = \frac{1}{n}M^{n+}(aq) + \frac{1}{2}H_2(g)$.

Thus, for copper, the ion is $Cu^{2+}(aq)$ and $n = 2$, so we have:

$$\frac{1}{2}Cu(s) + H^+(aq) = \frac{1}{2}Cu^{2+}(aq) + \frac{1}{2}H_2(g); \ \Delta G_m^{\ominus} = 32.7\,kJ\,mol^{-1} \qquad (21.1)$$

In Table 12.1, the metals with the most negative values of $\Delta G_m^{\ominus}$ were placed at the top, and those with the most positive values at the bottom. The strengths of the metals as reducing agents then decrease as you descend the Table.

When assembling Table 12.1, we emphasized the method by which $\Delta G_m^{\ominus}$ is obtained by combining the separate $\Delta H_m^{\ominus}$ and $\Delta S_m^{\ominus}$ values as ($\Delta H_m^{\ominus} - T\Delta S_m^{\ominus}$). However, *in certain cases*, $\Delta G_m^{\ominus}$ for reactions such as Reaction 21.1 can be obtained directly by electrical measurements. Consider Figure 21.1. The left-hand beaker contains a platinum metal electrode in contact with hydrogen gas and $H^+(aq)$. The right-hand beaker contains a copper metal electrode in contact with $Cu^{2+}(aq)$. The two beakers are connected by a glass tube containing a conducting solution of KCl(aq); this is known as a *salt bridge*.

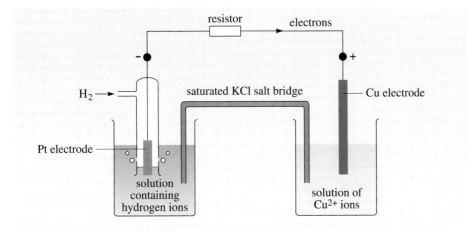

Figure 21.1 An electrochemical cell in which Reaction 21.2 occurs spontaneously.

This arrangement creates a voltage between the electrodes. If, as in Figure 21.1, the terminals are joined by a wire and small resistor, electrons flow in the wire from left to right. The electrode from which the electrons flow is marked negative; the electrode towards which they flow is marked positive. This electron flow is provided by reactions that occur at each electrode. These are shown in Table 21.1.

Table 21.1 Electrode reactions in an electrochemical cell

left electrode:	$H_2(g) = 2H^+(aq) + 2e^-$
right electrode:	$Cu^{2+}(aq) + 2e^- = Cu(s)$
overall reaction	$Cu^{2+}(aq) + H_2(g) = Cu(s) + 2H^+(aq)$ (21.2)

The arrangement shown in Figure 21.1 is called an **electrochemical cell**. It is shown in **discharge mode**; that is to say, it is releasing energy or doing electrical work by pushing an electric current through the resistor. The energy needed to do this comes from the overall chemical reaction that takes place in the cell. You exploit the same principle every day when you use electrical batteries (Figure 21.2), although these develop much higher voltages than the cell shown in Figure 21.1. This is discussed further in the *Batteries Case Study*.

Figure 21.2
Some examples of batteries containing different chemicals. In discharge mode, these chemicals react and supply energy.

The Gibbs energy change of this cell reaction can be determined from the voltage developed by the cell. This voltage can be found by replacing the resistor with a counter-voltage that can be adjusted to stop the natural flow of electrons in Figure 21.1. This 'stopping voltage' is equal to the maximum possible voltage that the cell can deliver. It is known as the **electromotive force (e.m.f.)**, *E*, of the cell. If we arrange things so that reactants and products in the total reaction are all in their standard states (Section 9.2), it is the **standard e.m.f.**, $E^\ominus$ of the cell. In the case of Figure 21.1, the stopping voltage when this is so is found to be 0.339 V. By convention, the e.m.f., in this case $E^\ominus$, is obtained by giving this stopping voltage the sign of the electrode on the right.

So what is the $E^\ominus$ value for the cell of Figure 21.1?

The electrode on the right is the copper electrode, which has a positive sign; $E^\ominus = +0.339$ V.

Now we must find the cell reaction to which this $E^\ominus$ value refers. The rule is that it must be written so that oxidation occurs at the left-hand electrode, and reduction at the right. In the case of Figure 21.1, that very reaction appears in Table 21.1: at the left-hand electrode, hydrogen is oxidized to hydrogen ions, and copper ions are reduced at the right. Thus, $E^\ominus = +0.339$ V refers to:

$$Cu^{2+}(aq) + H_2(g) = Cu(s) + 2H^+(aq); E^\ominus = 0.339 \text{ V} \qquad (21.3)$$

Now that the cell reaction has been properly written down, we can find its value of $\Delta G_m^\ominus$ from an equation that we state without proof:

$$\Delta G_m^\ominus = -nFE^\ominus \qquad (21.4)$$

Here, *n* is the number of electrons that are gained by the oxidized state at the left (Cu^{2+}) as it changes to the reduced state on the right (Cu). In this case, therefore, $n = 2$. *F* is a constant called the Faraday constant, whose value is 96.485 kJ mol^{-1} V^{-1}. Therefore

$$\Delta G_m^\ominus = -2 \times 96.485 \times 0.339 \text{ kJ mol}^{-1} \qquad (21.5)$$
$$= -65.4 \text{ kJ mol}^{-1}$$

Thus, for the cell reaction (Equation 21.3), we find

$$Cu^{2+}(aq) + H_2(g) = Cu(s) + 2H^+(aq); \Delta G_m^\ominus = -65.4 \text{ kJ mol}^{-1} \qquad (21.6)$$

⬤ Is this value consistent with that taken from Table 12.1 and shown in Equation 21.1?

⬤ Yes; if Equation 21.6 is reversed and halved, it becomes

$$\tfrac{1}{2}Cu(s) + H^+(aq) = \tfrac{1}{2}Cu^{2+}(aq) + \tfrac{1}{2}H_2(g); \quad \Delta G_m^{\ominus} = 32.7 \text{ kJ mol}^{-1} \qquad (21.1)$$

So far we have only described the measurement of the standard e.m.f. of the cell in which the right-hand beaker of Figure 21.1 contains copper metal and $Cu^{2+}(aq)$; see Equation 21.3. But clearly, we could obtain the standard e.m.f. of other cell reactions of the same type by replacing $Cu(s)$ and $Cu^{2+}(aq)$ by a different metal and its aqueous ion; for example

$$Ag^+(aq) + \tfrac{1}{2}H_2(g) = Ag(s) + H^+(aq); \quad E^{\ominus} = 0.80 \text{ V} \qquad (21.7)$$

$$Zn^{2+}(aq) + H_2(g) = Zn(s) + 2H^+(aq); \quad E^{\ominus} = -0.76 \text{ V} \qquad (21.8)$$

To abbreviate cell reactions 21.3, 21.7 and 21.8, we now replace each combination of $\tfrac{1}{2}H_2(g)$ on the left with $H^+(aq)$ on the right by the symbol e. This gives:

$$Cu^{2+}(aq) + 2e = Cu(s); \quad E^{\ominus} = 0.34 \text{ V} \qquad (21.9)$$

$$Ag^+(aq) + e = Ag(s); \quad E^{\ominus} = 0.80 \text{ V} \qquad (21.10)$$

$$Zn^{2+}(aq) + 2e = Zn(s); \quad E^{\ominus} = -0.76 \text{ V} \qquad (21.11)$$

These $E^{\ominus}$ values are called **standard redox potentials**. Table 21.2 contains a fuller list arranged in ascending order of $E^{\ominus}$.

There are four important points to note about this Table:

1 The symbol 'e' in each equation is not strictly an electron: if it were, it would have a superscript minus as in e^-. The symbol e is shorthand for $\tfrac{1}{2}H_2(g)$ on the side on which it appears, and $H^+(aq)$ on the other side. However, if you think of it as an electron, each reaction in the table becomes balanced with respect to charge.

2 In all of the reactions, the e symbol appears along with an oxidized state or oxidizing agent (e.g. Mg^{2+}) on the left; a reduced state or reducing agent (e.g. Mg) appears on the right. Thus a concise way of expressing equations such as 21.11 is to write $E^{\ominus}(Zn^{2+}|Zn) = -0.76 \text{ V}$, with the oxidized state first inside the bracket, and the reduced state second, with a vertical bar separating the two states. $\Delta G_m^{\ominus}$ for these equations can then be obtained from Equation 21.4.

3 The more powerful the reducing agent in a redox system, the more negative is the $E^{\ominus}$ value. Roughly speaking, redox systems with $E^{\ominus}$ values more negative than about -0.1 V contain powerful reducing agents and weak oxidizing agents; redox systems with $E^{\ominus}$ values more positive than about 1.1 V contain powerful oxidizing agents and weak reducing agents.

4 Redox potentials can be used to predict whether or not a particular redox reaction is thermodynamically possible. Because the system with the more negative $E^{\ominus}$ value contains the stronger reducing agent, we can make the following generalization:

> If we have two redox systems, and the $E^{\ominus}$ value for the first is more negative than that for the second, then the reduced state in the first is thermodynamically capable of reducing the oxidized state in the second.

Table 21.2 Standard redox potentials at 298.15 K

Redox reaction	$E^{\ominus}$/V
$Li^+(aq) + e = Li(s)$	−3.04
$Cs^+(aq) + e = Cs(s)$	−3.03
$Ca^{2+}(aq) + 2e = Ca(s)$	− 2.87
$Na^+(aq) + e = Na(s)$	−2.71
$Mg^{2+}(aq) + 2e = Mg(s)$	−2.36
$Zn^{2+}(aq) + 2e = Zn(s)$	−0.76
$Sn^{2+}(aq) + 2e = Sn(s)$	−0.14
$H^+(aq) + e = \tfrac{1}{2}H_2(g)$	0.00
$Cu^{2+}(aq) + 2e = Cu(s)$	0.34
$\tfrac{1}{2}I_2(s) + e = I^-(aq)$	0.53
$Fe^{3+}(aq) + e = Fe^{2+}(aq)$	0.77
$Ag^+(aq) + e = Ag(s)$	0.80
$\tfrac{1}{2}Br_2(aq) + e = Br^-(aq)$	1.10
$\tfrac{1}{2}Cl_2(g) + e = Cl^-(aq)$	1.36
$\tfrac{1}{2}F_2(g) + e = F^-(aq)$	2.89

Electrochemical cells are a very convenient way of determining $\Delta G_m^{\ominus}$ values in aqueous solution, but their application is limited. Often cells cannot be persuaded to develop their full e.m.f., and some electrode systems, like those of the alkali metals, are not viable because one of their components reacts with water. The $E^{\ominus}$ value can then be found by the non-electrical methods described earlier in this Book: $\Delta G_m^{\ominus}$ for the redox reaction is determined from $\Delta H_m^{\ominus}$ and $\Delta S_m^{\ominus}$, and then $E^{\ominus}$ is calculated from Equation 21.4. Question 21.1 provides an example of this.

QUESTION 21.1

Aluminium is one of the metals whose standard redox potential cannot be determined by an electrochemical measurement. In the *Data Book*, $\Delta G_f^{\ominus}(\text{Al}^{3+}, \text{aq}) = -485 \text{ kJ mol}^{-1}$. This is the value of $\Delta G_m^{\ominus}$ for the following reaction:

$$\text{Al(s)} + 3\text{H}^+(\text{aq}) = \text{Al}^{3+}(\text{aq}) + \tfrac{3}{2}\text{H}_2(\text{g}); \; \Delta G_m^{\ominus} = -485 \text{ kJ mol}^{-1} \qquad (21.12)$$

Use this to obtain a value of $\Delta G_m^{\ominus}$ for the process written as

$$\text{Al}^{3+}(\text{aq}) + 3e = \text{Al(s)} \qquad (21.13)$$

and then use your value to calculate $E^{\ominus}(\text{Al}^{3+}|\text{Al})$. Does your result suggest that aluminium is a strong reducing agent?

QUESTION 21.2

(a) What are the two strongest and two weakest oxidizing agents in Table 21.2?

(b) What are the two strongest and two weakest reducing agents in Table 21.2?

QUESTION 21.3

Use the information in Table 21.2 to predict which of the following reactions are thermodynamically favourable:

(i)   reduction of $\text{Zn}^{2+}(\text{aq})$ by magnesium metal;

(ii) reduction of $\text{Sn}^{2+}(\text{aq})$ by zinc metal;

(iii) reduction of $\text{Ca}^{2+}(\text{aq})$ by zinc metal;

(iv) oxidation of $\text{Fe}^{2+}(\text{aq})$ by $\text{Cl}_2(\text{g})$;

(v) oxidation of $\text{Br}^-(\text{aq})$ by $\text{Cl}_2(\text{g})$.

Where a reaction occurs, write an equation for it.

IONIZATION ENERGIES OF ATOMS

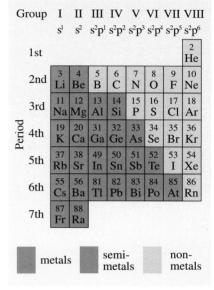

22

In Section 19, we promised that Sections 20–22 would deal with three thermo-dynamic quantities. The lattice energy (Section 20) drew attention to ions in the solid state; the redox potential (Section 21) focused on ions in aqueous solution. Our third quantity, the ionization energy, is concerned with *gaseous* ions and their formation from *gaseous* atoms.

In Section 18, the ionization energy of a chemical species A was defined as the molar enthalpy change at 298.15 K, for the reaction:

$$A(g) = A^+(g) + e^-(g) \tag{22.1}$$

In other words, it is the molar enthalpy change for the removal of an electron from A. Figure 22.2 shows how the first ionization energies of the atoms in Figure 22.1 (and hydrogen) vary with increasing atomic number.

- What is the general nature of the change that occurs across each Period from Group I to Group VIII?

- There is an overall increase in ionization energy. Compare, for example, the values for lithium and neon.

Let's try to relate this change to the change in electronic configuration. At helium, the electronic configuration of the atom is $1s^2$, and at lithium it is $1s^22s^1$. We can therefore write the latter as $[He]2s^1$, where $[He]$ denotes the helium configuration. At neon, at the right-hand end of Period 2, the configuration is $[He]2s^22p^6$. Thus,

Figure 22.1
The mini-Periodic Table, showing the typical elements.

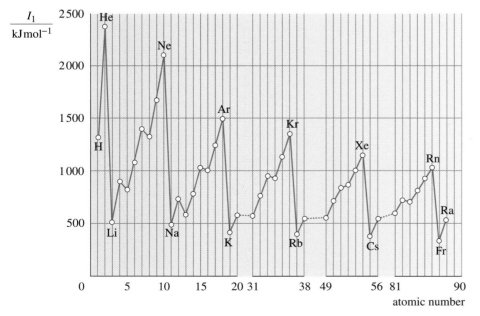

Figure 22.2 The first ionization energies, I_1, of the elements in Figure 22.1, plotted against atomic number. Note the breaks on the horizontal axes where the transition series and the lanthanides have been omitted.

in moving across the Period from lithium to neon, one proton is added to the nucleus at each step, and one electron is added to the shell with principal quantum number $n = 2$. Let's consider these two effects in turn. Concentrate first on just one of the electrons in the shell with $n = 2$, and imagine that this is the one that is removed on ionization.

⬤ What overall effect should the successive addition of protons have on the ionization energy from lithium to neon?

⬤ It should produce an overall increase: the build up of positive charge on the nucleus means that the electron is subjected to an increasingly strong attractive force; this suggests that progressively more energy will be needed to remove the electron from the atom.

Now we turn to the second effect.

⬤ What overall effect should the successive addition of electrons to the shell with $n = 2$ have on the ionization energy from lithium to neon?

⬤ It should produce an overall decrease: the build up of accompanying electrons in the shell means that the original electron is subjected to an increasingly strong repulsive force; this tends to make ionization progressively easier.

Thus, the two effects compete with each other: one tends to produce an overall increase across the Period, and the other an overall decrease.

⬤ Which effect turns out to be dominant?

⬤ The increase in the nuclear charge: the experimental results in Figure 22.2 show that this effect overwhelms the influence of electron repulsion to produce an overall increase in ionization energy across the Period.

This dominance of the increasing nuclear charge is perhaps not surprising since it acts from the centre, whereas the increasing electron repulsion arises from electrons that are dispersed in shells around the nucleus.

Having accounted for the overall increase, let's now turn to the deviations that occur from it. Figure 22.3 shows the variation from lithium to neon in an enlarged form. The upward trend is broken by downward displacements between beryllium and boron, and between nitrogen and oxygen.

⬤ How would you relate the break between beryllium and boron to the change in electronic configuration?

⬤ It marks the point where the electrons stop entering an s sub-shell, and begin filling up a p sub-shell.

In explaining the overall increase in ionization energy between lithium and neon, we assumed that, as the electron that is lost is always in the shell with $n = 2$, it is in the same state at each element. But this is not so: for the elements lithium and beryllium, the electron being lost is in a 2s sub-shell; for the elements boron to neon, it is in a 2p sub-shell. It is therefore not surprising that the plot of the ionization energies for the second set of elements is displaced with respect to that for

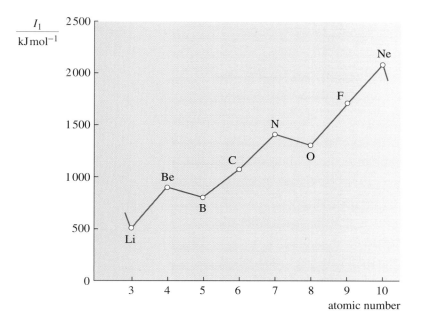

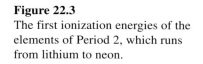

Figure 22.3
The first ionization energies of the elements of Period 2, which runs from lithium to neon.

the first. As the displacement is downwards, the implication is that, other things being equal, there is something about s electrons that makes them harder to remove than p electrons. This point is taken further in *Elements of the p Block*[3].

The second downward break, between nitrogen and oxygen, can be related to Hund's rule.

● Sketch the box diagrams for the electronic ground states of the elements boron to neon. How do oxygen, fluorine and neon differ from boron, carbon and nitrogen?

● The 2p sub-shells in the box diagrams are shown overleaf (Figure 22.4). Unlike those of boron, carbon and nitrogen, the 2p sub-shells of oxygen, fluorine and neon contain paired electrons. On ionization, it is one of these paired electrons that is lost.

Thus, the ionization of an oxygen, fluorine or neon atom leads to a reduction in the number of paired electrons, whereas the ionization of a boron, carbon or nitrogen atom does not. Now Hund's rule tells us that the ground state of an atom is the one that maximizes the number of parallel spins in an unfilled sub-shell. This marks the fact that pairs of electrons with opposed spins in a sub-shell repel one another more strongly than do pairs with parallel spins. Thus, an ionization that eliminates paired spins is favoured relative to one that does not. Consequently, in Figure 22.2, the plot for O, F and Ne is displaced downwards with respect to that for B, C and N.

If you examine Figure 22.2, you will see that the breaks between Groups II and III, and between Groups V and VI, appear in some later Periods as well, although in a less marked form. Our argument shows that those breaks separate atoms that contain np electrons from those that contain just ns electrons, and atoms that contain p electrons with opposed spins from those that do not.

22.1 Changes in ionization energy down a Group

You have just seen how the increase in nuclear charge leads to an overall increase in ionization energy between lithium and neon. If the 2p sub-shell could contain more than six electrons, then one would expect sodium, the element that follows neon, to have a higher ionization energy than any element in the lithium–neon Period. However, as Figure 22.2 shows, there is a steep drop after neon; indeed, it is so steep that the ionization energy of sodium is lower than that of lithium, or indeed any other element in Period 2.

⬤ The electronic configurations of the lithium and sodium atoms are [He]$2s^1$ and [Ne]$3s^1$, respectively. Why, despite its much higher nuclear charge, does the sodium atom have the lower ionization energy?

⬤ The sodium electron lost on ionization is in the shell with $n = 3$. The electrons in this shell tend to be much further from the nucleus than those in the shell with $n = 2$. This, presumably, is the reason why the outer electron in sodium is held less strongly than that in lithium, even though the sodium atom has the greater nuclear charge.

This decrease in ionization energy between lithium, and sodium, the element below it in the same Group, illustrates a general trend. The ionization energies of the typical elements tend to decrease down a Group as the outer electrons occupy shells of higher principal quantum number which are further from the nucleus. The tendency is discernible in Figure 22.2: compare the values for the halogens and noble gases. It is even clearer in Figure 22.5, which also reveals a small number of exceptions. Although the sizes of these exceptions look small, their consequences are quite important; they are taken up in *Elements of the p Block*[3].

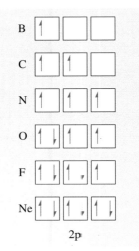

Figure 22.4
Box diagrams for the electronic structures of the atoms boron to neon.

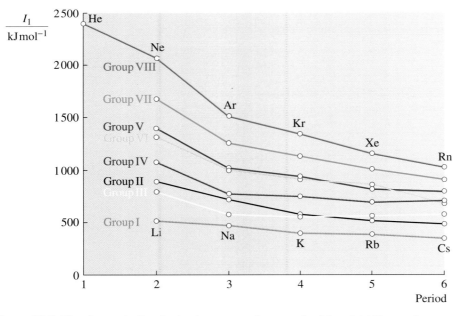

Figure 22.5 The change in first ionization energy down each of the eight Groups in Figure 22.1.

22.2 Summary of Section 22

1   The first ionization energies of the atoms of the typical elements show an *overall* increase across each Period, and an *overall* decrease down each Group. The overall increase across each Period is broken at points where p electrons first join s electrons in the outer electronic configuration, and where p electrons in the box diagram can no longer all have parallel spins.

2   The overall increase in ionization energy across each Period shows the importance of the effect of increasing nuclear charge. This binds the outer electrons more and more tightly across each row.

3   The overall decrease in ionization energy down each Group, shows the importance of the increase in distance of the outer electrons from the nucleus as the successive electron shells build up. This increasing distance weakens the attractive force between the nucleus and the outer electrons.

QUESTION 22.1

By referring to Figure 22.1 only, state which atom of each of the following pairs of elements has the higher first ionization energy: (i) aluminium and sulfur; (ii) nitrogen and antimony; (iii) germanium and chlorine.

THE CHEMISTRY OF GROUP I: THE ALKALI METALS

23

We now begin at the descriptive chemistry of the typical elements. The rest of this Book is taken up with Groups I and II (the other Groups are dealt with in *Elements of the p Block*[3]). Here, we start with Group I, the alkali metals. Figure 23.1 shows their position in the Periodic Table. Most of the compounds that they form are usually regarded as ionic, and treated by an ionic model. In Sections 18–22 we discussed Born–Haber cycles, lattice energies and ionization energies. These concepts are highly relevant to the ionic model and will therefore prove useful in understanding alkali metal chemistry.

Of the five common alkali elements, lithium, rubidium and caesium are all much less common than sodium or potassium, which are, respectively, the seventh and eighth most abundant elements in the Earth's crust. In many parts of the world, there are vast deposits of sodium chloride, which have been formed by the evaporation of ancient seas (Figure 23.2). At present, world industry consumes about 190 million tonnes of sodium chloride each year, but this is merely a drop in the dried-out ocean of the Earth's total reserves. The UK's source in Cheshire contains more than 100 000 million tonnes, and much larger deposits occur elsewhere.

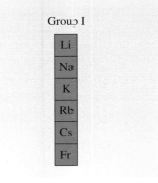

Figure 23.1
The alkali metals in the Periodic Table.

Figure 23.2 The Wieliczca salt mines near Cracow in Poland include an underground chapel with chandeliers made from crystalline salt.

23.1 The alkali metals

The alkali metals occur near the top of Tables 12.1, 15.1, 15.2 and 21.2. This shows that, thermodynamically speaking, they have a strong tendency to be oxidized.

● What does this suggest about the difficulty of extracting such metals from their compounds?

● It will be considerable; if the metals are readily oxidized, the resulting ions or compounds will be difficult to reduce.

In Section 17.3, you saw that metals of this type are obtained, either by reducing their compounds with another highly oxidizable metal such as calcium, or by electrolysis. Therefore it is not surprising that lithium and sodium are obtained by electrolysis of their molten chlorides, and that rubidium and caesium are made by the reduction of their chlorides by calcium metal.

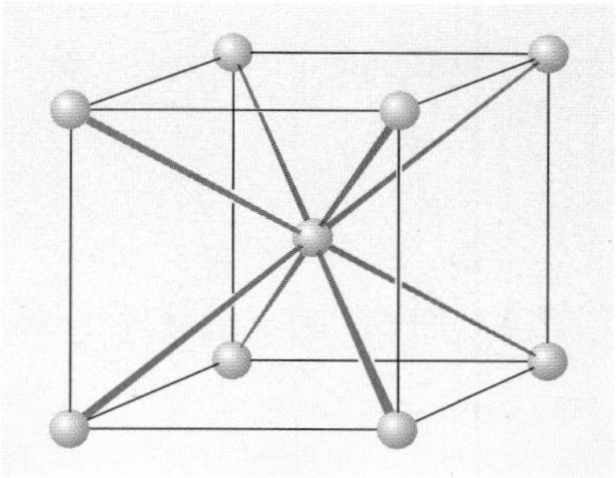

The five alkali metals have the body-centred cubic structure (Figure 23.3). Some physical properties of the metals, their atoms and their ions are shown in Table 23.1.

Figure 23.3
Metallic lithium, sodium, potassium, rubidium and caesium have a body-centred cubic structure.

Table 23.1 Some properties of the alkali metal atoms and ions

	Li	Na	K	Rb	Cs
electronic configuration	[He]2s^1	[Ne]3s^1	[Ar]4s^1	[Kr]5s^1	[Xe]6s^1
I_1/kJ mol^{-1}	520	496	419	403	376
I_2/kJ mol^{-1}	7 298	4 562	3 051	2 632	2 236
$r(M^+)$/pm	90	116	152	166	181
$E(M)$/kJ mol^{-1}	60	53	48	47	46
$\Delta H_{atm}^{\ominus}(M, s)$/kJ mol^{-1}	159	107	89	81	76
melting temperature/°C	181	98	63	39	29
boiling temperature	1 347	881	766	688	705
density/g cm^{-3}	0.53	0.97	0.86	1.53	1.90
$E^{\ominus}(M^+ \vert M)$/V	−3.04	−2.71	−2.94	−2.94	−3.03

DATA BOOK

Pg 18, SECTION 6

} Pg 34, TABLE 10.1

IONIC RADII, Pg 38, SECTION 14

Pg 36, SECTION 11, ELECTRON AFFINITY

Pg 21, TABLE 8.1

The alkali metals have unusually low densities, melting temperatures and boiling temperatures. For example, lithium, sodium and potassium are less dense than water: during their vigorous reactions with water, they float (Figure 23.4).

In Section 1.1 we discussed the simple electron gas model of metallic bonding; this is pictured again in Figure 23.5. The atoms of the metallic elements form positive ions and free electrons, and the latter are dispersed throughout the structure. The dispersed electrons act as a sort of glue: their attraction for the oppositely charged positive ions holds the ions together in the metallic structure.

● Use this model to suggest why the alkali metals have low melting temperatures and low densities.

Figure 23.4
The low density of sodium is revealed by the fact that it floats during its reaction with water.

149

The alkali metals form cations with a single positive charge. Thus, the attraction between the electron gas and the ion cores is weaker than in other metals, and the positive ions are not pulled together so strongly. The densities and melting temperatures are therefore lower.

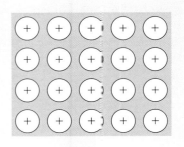

To reinforce this point, Table 23.2 compares the relevant properties of sodium with those of magnesium and aluminium, the two adjacent metals in the Periodic Table (Figure 23.6).

Figure 23.5
The electron gas model of metallic bonding.

Table 23.2 Some properties of metallic sodium, magnesium and aluminium, and of their atoms

	Na	Mg	Al
electronic configuration	$[Ne]3s^1$	$[Ne]3s^2$	$[Ne]3s^23p^1$
$\Delta H_{atm}^{\ominus}$(M, s)/kJ mol^{-1}	107	148	326
I_1/kJ mol^{-1}	496	738	578
I_2/kJ mol^{-1}	4562	1451	1817
density/g cm^{-3}	0.97	1.74	2.70
melting temperature/°C	98	649	660

Using the electron gas model, what are the expected charges on the ion cores in metallic magnesium and aluminium?

They are +2 and +3, respectively. As with Na$^+$, there is a tendency to form ions with a noble gas configuration (neon); these ions are Mg^{2+} and Al^{3+}.

With these charges, the increase in density and melting temperature from sodium through magnesium to aluminium is what we would expect. Another mark of the unusually weak bonding in the alkali metals is the low value of $\Delta H_{atm}^{\ominus}$(M, s), the standard enthalpy change of the reaction:

$$M(s) = M(g) \tag{23.1}$$

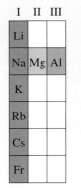

Figure 23.6
The relative positions of the alkali metals, magnesium and aluminium in the Periodic Table of Figure 22.1.

Table 23.2 also shows that the first ionization energy of sodium is lower than that of magnesium or aluminium; considering the family as a whole, the Group I elements have lower first ionization energies than any other Group in the Periodic Table. As you saw in Section 22, this is because the nuclear charge that acts on a particular electron shell is lowest at the beginning of the Period, where the Group I elements occur. Now, when the alkali metals form free M$^+$ ions, the metal is first broken down into atoms (Equation 23.1), and then the gaseous atom is ionized:

$$M(g) = M^+(g) + e^-(g) \tag{23.2}$$

The relatively low enthalpy changes for Equations 23.1 and 23.2 therefore show that the alkali metals are more readily converted into *gaseous* M$^+$ ions than any other group of metals. In Section 18 we showed how the low values of $\Delta H_{atm}^{\ominus}$(M, s) and I_1(M) also accounted for the readiness of the metals to be oxidized to *compounds* or *solutions* containing M$^+$ ions; for example

$$M(s) + \tfrac{1}{2}Cl_2(g) = MCl(s) \tag{23.3}$$

$$M(s) + H_2O(l) = M^+(aq) + HO^-(aq) + \tfrac{1}{2}H_2(g) \tag{23.4}$$

In this case, readiness to be oxidized was understood in a thermodynamic sense: the low values of $\Delta H_{atm}^{\ominus}(M, s)$ and $I_1(M)$ tend to make Reactions 23.3 and 23.4 thermodynamically very favourable. However, this by itself cannot account for the explosive nature of such reactions; a second, complementary condition is required.

⬤ What condition is this?

⬤ A kinetic one: the reactions must not only be thermodynamically very favourable, but they must also be *fast*.

The reactions of the alkali metals tend to be fast because of the low values of $\Delta H_{atm}^{\ominus}(M, s)$. The rate of the reaction is limited by the speed with which the metal structure can be broken down; a metal with a low value of $\Delta H_{atm}^{\ominus}(M, s)$ is easily taken apart.

⬤ If this is so, use Table 23.1 to decide which alkali metal should react most slowly.

⬤ The values of $\Delta H_{atm}^{\ominus}(M, s)$ decrease down the Group; lithium has the largest value and should therefore react most slowly.

As you know, when the alkali metals are dropped into water, lithium undergoes a relatively mild reaction.

Although the alkali metals readily form compounds or solutions containing M^+ ions, the ionization never goes further to give substances containing M^{2+} ions. Tables 23.1 and 23.2 show why: the second ionization energies of the alkali metals are enormous — much higher than those of magnesium or aluminium.

⬤ Why is I_2 for an alkali metal such as sodium so large?

⬤ The second ionization energy, I_2, is the ionization energy of the ion, Na^+. This ion has the noble gas structure of the neon atom. On ionization, an electron must be removed from an inner shell (principal quantum number $n = 2$) of the sodium atom whose electrons are much closer to the nuclear charge.

The very low first ionization energies and very high second ionization energies of the alkali metals are a quantitative indication of the stability of the noble gas electronic configuration, which plays such an important role in the elementary bonding theories that you should be familiar with. The low first ionization energies suggest that the noble gas configuration is *relatively* easily formed from the metals; the high second ionization energies show that it is broken into only with great difficulty.

23.1.1 The alkali metals in liquid ammonia

Table 23.1 contains the standard redox potentials of the alkali metals in water, $E^{\ominus}(M^+|M)$. These are all large and negative. They put the alkali metals at the top of Table 21.2, and show that they are very powerful reducing agents in aqueous solution. Here we use a very remarkable example to show that this reducing power is evident in other solvents also. Ammonia gas, NH_3, like water, is the hydride of a non-metal. It can be liquefied by cooling it below its boiling temperature of $-33\,°C$. Like water, the liquid is colourless. If an alkali metal is added to liquid ammonia in small amounts, the metal dissolves and a blue solution is formed (Figure 23.7).

The alkali metal forms colourless positive ions, M^+(amm), where 'amm' signifies that they are dissolved in ammonia; the blue colour is due to electrons, which like the positive ions, are dispersed throughout the ammonia solvent:

$$M(s) = M^+(amm) + e^-(amm) \qquad (23.5)$$

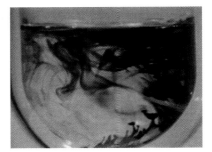

● What has happened to the alkali metal in this reaction?

● It has undergone oxidation; it has lost an electron.

● How does Reaction 23.5 differ from a typical redox reaction?

● The alkali metal has been oxidized, but nothing seems to have been reduced. Oxidation and reduction are usually complementary processes.

Figure 23.7
The blue colour of ammoniated electrons developing as sodium dissolves in liquid ammonia.

What has happened here is that the electrons that have been lost have not been passed to an oxidizing agent. They have been 'parked' in the solvent, where they are awaiting something that they can reduce. Not surprisingly, therefore, solutions of the alkali metals in liquid ammonia find many uses as strong reducing agents in both organic and inorganic chemistry. Reactions of type 23.5 are rare; apart from the alkali metals, the only metals that undergo them are the Group II elements, calcium, strontium and barium, and two of the lanthanide elements, europium and ytterbium. These, like the alkali metals, are all metals with a high tendency to be oxidized.

23.2 Summary of Section 23

1   The alkali metals have low melting temperatures and boiling temperatures, and are very easily oxidized. They react readily with water giving alkaline solutions of hydroxides and hydrogen gas. The vigour of this reaction increases down the Group. The metals also dissolve in liquid ammonia, giving blue solutions containing the ammoniated electron. When the alkali metals are oxidized in these and other reactions, compounds or solutions containing M^+ ions are formed.

2   In alkali metal atoms, the nuclear charge is lower than in other elements in the same Period. The first ionization energies, in which a noble gas configuration is formed, are therefore relatively low. By contrast, the second ionization energies are unusually large because an electron must be removed from an inner shell of the atom where electrons are closer to the nucleus than the outer s^1 electron.

3   The combination of low first, and high second ionization energies is primarily responsible for the readiness of the alkali metals to form substances containing M^+ ions, but not ions of any higher charge.

4   According to the simple electron-gas model of metallic bonding, the alkali metals contain only one freed electron per atom. Thus, the interaction between the ion core and electron gas is relatively weak. This accounts for the softness, and for the low densities, melting temperatures and atomization energies of the alkali metals.

5   Because the value of $\Delta H_{atm}^{\ominus}(M, s)$ is low, the structure of an alkali metal is easily broken down and the metals tend to react relatively quickly.

QUESTION 23.1

The values of $\Delta H_m^{\ominus}$ and $\Delta G_m^{\ominus}$ for the reactions of lithium and caesium with water are very similar. However, caesium reacts explosively and lithium steadily; explain the difference in terms of the properties of the atoms and their ions.

ALKALI METAL COMPOUNDS IN INDUSTRY

24

Most alkali metal compounds are quite soluble in water. This is especially the case with compounds of sodium, the most common of the five elements. When the compounds dissolve, their positively charged ions, such as $Na^+(aq)$, are formed, and these are very hard to reduce, and impossible to oxidize. Thus one of the most important uses of sodium is as an inert cation (the 'counter ion') that can carry negatively charged ions into solution. For example, few carbonates or hydroxides, apart from those of the alkali metals, dissolve easily in water, yet solutions containing the ions $OH^-(aq)$ or $CO_3^{2-}(aq)$ have many practical uses both in industry and elsewhere. Thus, sodium carbonate, Na_2CO_3, and sodium hydroxide, $NaOH$, are produced by the chemical industry in very large amounts (Figure 24.1). Their value as sources of hydroxide and carbonate ions is enhanced by the resistance of the accompanying sodium ions to oxidation or reduction: there is no chance that chemicals that are meant to react with hydroxide or carbonate will react with the sodium ions instead.

Figure 24.1 The undistinguished appearance of sodium carbonate (left) and sodium hydroxide (right) gives no hint of their huge industrial importance.

The importance of sodium carbonate and sodium hydroxide is apparent from the fact that about 60% of the sodium chloride consumed in chemical manufacturing worldwide is used in a sector known as the **chlor-alkali industry**. The major products of this sector are chlorine, sodium hydroxide and sodium carbonate. To give you a flavour of the industry we shall simply describe the most modern method of making two of its three chief products: sodium hydroxide and chlorine.

24.1 Manufacture of sodium hydroxide and chlorine

Sodium hydroxide and chlorine are obtained commercially by the electrolysis of concentrated sodium chloride solutions. It sounds simple, but there is an important problem that must be solved first. To appreciate this, look first at the simplest possible arrangement, in which two electrodes are immersed in a concentrated solution of sodium chloride, and connected to a battery (Figure 24.2).

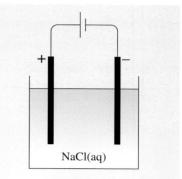

Figure 24.2
The simplest possible way of electrolysing a concentrated solution of sodium chloride.

⚫ Write down an equation for what happens at the positive electrode.

⚫ The equation is:

$$Cl^-(aq) = \tfrac{1}{2}Cl_2(g) + e^- \qquad (24.1)$$

Negative chloride ions in the solution migrate to the positive electrode, surrender their electrons, and are then evolved as chlorine gas.

Now for the negative electrode. An over-hasty judgement might be that here the positive sodium ions might be discharged as sodium metal:

$$Na^+(aq) + e^- = Na(s) \qquad (24.2)$$

This is incorrect, but by starting in this way we can arrive at the correct reaction.

⚫ Suppose that sodium metal were formed in this way. Write an equation for what would happen to it.

⚫ The equation is:

$$Na(s) + H_2O(l) = Na^+(aq) + OH^-(aq) + \tfrac{1}{2}H_2(g) \qquad (24.3)$$

Because water is present, sodium metal will react with it in the usual fashion. Thus, the overall process at the negative electrode would be the sum of Equations 24.2 and 24.3.

⚫ Add them up, and write down their sum.

⚫ Both the sodium ions, and sodium metal cancel out, leaving

$$H_2O(l) + e^- = OH^-(aq) + \tfrac{1}{2}H_2(g) \qquad (24.4)$$

To summarize, the initial reactions during the electrolysis of sodium chloride are as follows:

positive electrode: $\qquad Cl^-(aq) = \tfrac{1}{2}Cl_2(g) + e^- \qquad (24.1)$

negative electrode: $\quad H_2O(l) + e^- = OH^-(aq) + \tfrac{1}{2}H_2(g) \qquad (24.4)$

Because the sodium ions in the solution remain unchanged, the build-up of hydroxide ions at the negative electrode amounts to a build-up of aqueous sodium hydroxide solution. Thus, the initial products are chlorine gas at the positive electrode and sodium hydroxide solution at the negative electrode. In principle, therefore, both our desired products can be obtained from one electrolytic cell. However, there is a danger that they will indulge in mutual destruction because hydroxide ions and chlorine gas react together to produce the chloride ion, $Cl^-(aq)$, and the hypochlorite ion, $ClO^-(aq)$:

$$Cl_2(g) + 2OH^-(aq) = Cl^-(aq) + ClO^-(aq) + H_2O(l) \qquad (24.5)$$

It is therefore essential that the products produced at the positive and negative electrodes should be kept separate. The chemical industry has devised some ingenious ways of doing this.

24.1.1 The membrane cell

The separation of the chlorine and sodium hydroxide can be achieved by a membrane. The membrane allows the passage of the positive ion, $Na^+(aq)$, but not of negative ions such as $Cl^-(aq)$ and $HO^-(aq)$. Figure 24.3 shows such a membrane cell. The positive electrode (left) is usually metallic titanium, often coated with ruthenium dioxide, RuO_2; the negative electrode (right) is woven steel wire. Chlorine is formed in the left-hand compartment.

If chloride ions could pass through the membrane from left to right, they would become unavailable for chlorine production and adulterate the sodium hydroxide solution that is formed around the negative electrode. The U-shaped arrow in the left compartment therefore marks these two valuable effects of the membrane.

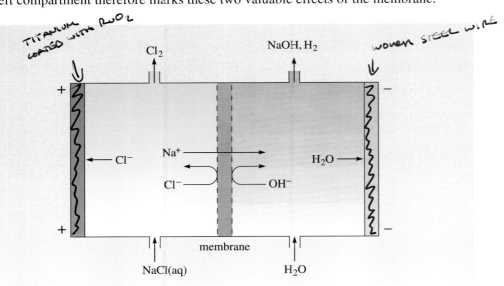

Figure 24.3 Schematic diagram of the manufacture of sodium hydroxide and chlorine, using a membrane cell. Ideally, the membrane allows passage of positive ions only. Hydrox-ide ions cannot get into the positive compartment of the cell to destroy chlorine, and chloride ions cannot get into the negative compartment to contaminate the sodium hydroxide solution.

● What valuable effect is marked by the U-shaped arrow in the right-hand compartment?

● The membrane stops $OH^-(aq)$ from entering the positive compartment and destroying chlorine via Equation 24.5.

What sort of material is the membrane made of? It is an organic polymer whose basic skeleton has been fluorinated so that it resembles polytetrafluoroethene, $(-CF_2)_n$, which is also known as PTFE or teflon. However, in this instance the structure of the skeleton is modified at some points by bridging the $(-CF_2-)$ groups with oxygen atoms, for example $(-CF_2-O-CF_2-)$. Groups of atoms are then attached to this basic skeleton at intervals in place of a fluorine atom. In what is known as the acid form of the polymer, these groups are $-COOH$ and $-SO_2OH$, and when the membrane makes contact with an aqueous solution, hydrogen ions tend to be lost; for example, where the acid form contains a carboxyl group, the following equilibrium will be present

$$F-C-C-O-H \text{ (mem)} = F-C-C-O^- \text{(mem)} + H^+ \text{(aq)} \qquad (24.6)$$

If the aqueous solution contains a high concentration of another positive ion such as Na^+(aq), the negatively charged sites so created can be occupied by this ion:

$$F-C-C-O^- \text{ (mem)} + Na^+ \text{(aq)} = F-C-C-O-Na \text{ (mem)} \qquad (24.7)$$

Suppose such occupation can occur throughout the structure. Then, if the concentration of sodium is higher on one side of the membrane than the other, sodium ions can pass through the membrane from one site to another to reduce the concentration difference (Figure 24.4).

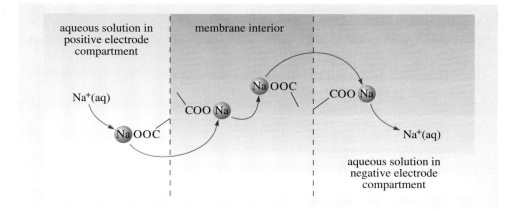

Figure 24.4 The membrane in a chlor-alkali membrane cell consists of a polymeric backbone based on polytetrafluoroethene. At intervals on this backbone there are $-COO^-$ and $-SO_2O^-$ sites, which can bind positive ions like Na^+ and H^+. Aqueous positive ions can therefore be transferred through the membrane from site to site, but aqueous negative ions cannot. For this reason, the membrane is known as a *cation-exchange membrane*.

⬤ Why can't negative ions such as Cl^-(aq) and OH^-(aq) enjoy the same experience?

⬤ The membrane can offer only negatively charged sites; there are no positive sites where negative ions might be bound.

Notice also that, because there is no bulk flow of water through the membrane, the water needed for the reaction at the negative electrode (Equation 24.4) cannot come from the brine added to the positive electrode compartment. So water is added separately to the negative electrode compartment. The brine concentration in the positive electrode compartment is maintained by combining a recycling of the depleted sodium chloride with a boost to its concentration (Figure 24.3). Solid sodium hydroxide is obtained by evaporation of the solution leaving the negative compartment.

24.2 Uses of sodium hydroxide and sodium carbonate

The uses of sodium hydroxide in industry are shown in Figure 24.5. As we have seen, the most striking property of alkali metal hydroxides is their solubility in water, and the resulting ability to deliver high concentrations of OH⁻(aq). Hydroxide ions destroy acidity, in the shape of H⁺(aq), through the neutralization reaction:

$$H^+(aq) + OH^-(aq) = H_2O(l) \qquad (24.8)$$

This explains the sector labelled 'neutralization' in Figure 24.5.

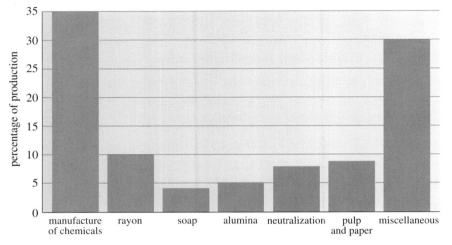
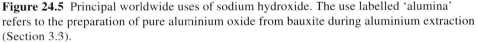

Figure 24.5 Principal worldwide uses of sodium hydroxide. The use labelled 'alumina' refers to the preparation of pure aluminium oxide from bauxite during aluminium extraction (Section 3.3).

World industry produces some 30 million tonnes of sodium hydroxide each year, along with a similar quantity of sodium carbonate, Na_2CO_3, the other sodium product of the chlor-alkali industry. A typical use-pattern of sodium carbonate is shown in Figure 24.6. As with sodium hydroxide, it is the alkalinity of the substance that explains its use in the manufacture of paper, soaps and detergents. We discuss its role in glass-making in Section 28.3.

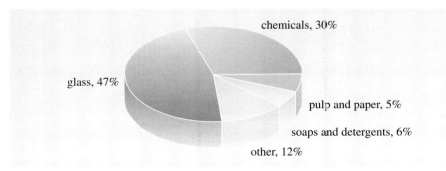

Figure 24.6 The principal worldwide uses of sodium carbonate

24.3 Summary of Section 24

1 The most important industrial compounds of the alkali metals are sodium hydroxide and sodium carbonate. Together with chlorine gas, these two substances are the major products of the chlor-alkali industry.

2 Sodium hydroxide is produced at the negative electrode, and chlorine at the positive electrode, when brine is electrolysed. The two products must be kept separate because chlorine and hydroxide ions react together. In membrane cells, there is a cation-exchange membrane, with sites that permit the passage of only positive ions. Brine is then cycled through the positive electrode compartment, and the negative compartment must be topped up with water.

BINARY ALKALI METAL COMPOUNDS WITH NON-METALS

25

25.1 Alkali metal halides

Alkali metal halides can be made by the direct reaction of the alkali metals and halogens. They are all white crystalline solids melting at high temperatures (450–1 000 °C) to give electrically conducting melts, and they all dissolve in water, yielding colourless solutions that conduct electricity; for example

$$NaI(s) = Na^+(aq) + I^-(aq) \tag{25.1}$$

All the alkali metal halides have formulae of the type MX, where M is an alkali metal atom, and X is a halogen atom. However, because of these properties, so characteristic of ionic compounds, they are taken to consist of M^+ and X^- ions.

Apart from CsCl, CsBr and CsI, all the alkali metal halides have the sodium chloride structure, whose unit cell is shown in Figure 25.1. Each sodium ion is octahedrally surrounded by six chloride ions. The compounds CsCl, CsBr and CsI have the caesium chloride structure (Figure 25.2), in which there are eight halides around each caesium.

The alkali metal halides are substances of great theoretical importance. This is because the ionic model of chemical bonding is developed by constant reference to, and assessment against, the properties of the alkali metal halides. For example, they can be used to generate a set of ionic radii that help us to understand why caesium chloride has the structure that it does. The larger cation in the caesium halides

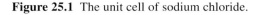

Na⁺
Cl⁻

Figure 25.1 The unit cell of sodium chloride.

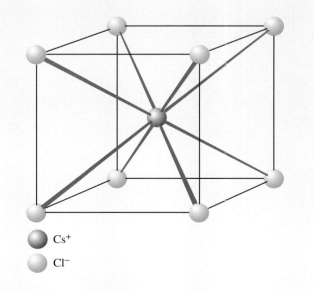

Cs⁺
Cl⁻

Figure 25.2 The unit cell of caesium chloride.

(see Table 23.1) provides the extra room needed to pack eight rather than six chloride, bromide or iodide ions around the metallic element.

Likewise, in Section 20, the alkali metal halides were very useful in testing our attempts to use an ionic model for the prediction of lattice energies.

25.2 Oxygen compounds of the alkali metals

When the alkali metals are heated in plenty of oxygen, they combine with it to form compounds. We might expect such compounds to contain M^+ and O^{2-} ions with the formula M_2O. In the case of lithium, this is quite correct. The oxide Li_2O is formed, and it has the antifluorite structure, which has the unit cell shown in Figure 25.3. We may think of it as an assembly of Li^+ and O^{2-} ions. Each oxide ion is surrounded by eight lithium ions at the corners of a cube; each lithium ion is tetrahedrally coordinated to four oxide ions.

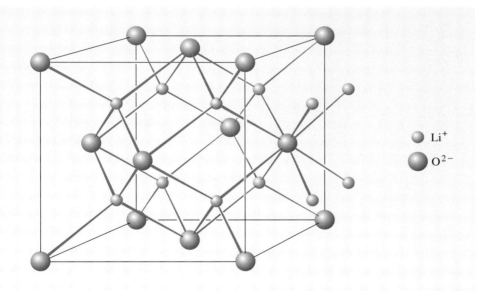

Li^+

O^{2-}

Figure 25.3 The unit cell of lithium oxide, Li_2O.

However, the compound formed by sodium is not Na_2O; it has the empirical formula NaO, where there is one oxygen for every sodium.

If this were a normal, ionic oxide, what would be the charge on the cation?

A normal ionic oxide is taken to contain O^{2-} ions; the cation would therefore be Na^{2+}.

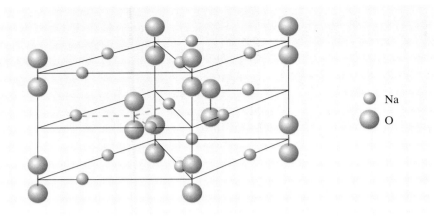

Na

O

Figure 25.4
The structure of sodium peroxide, Na_2O_2: the oxygens are grouped in pairs with an internuclear distance of 149 pm, so the compound is assumed to contain peroxide ions with the formula O_2^{2-}.

Now, we said in Section 23.1 that compounds containing Na^{2+} ions are precluded by the very high second ionization energy of sodium. But this inconsistency is only apparent because the compound of empirical formula NaO is *not* a normal oxide; this is made clear by the crystal structure, which has the unit cell shown in Figure 25.4. In normal oxides, individual oxygen atoms are bound only to the atoms of other elements, as in Figure 25.3. In NaO, the oxygens are bound in pairs, and the distance between them is 149 pm, not much more than the distance 121 pm found in the O_2 molecules of oxygen gas. The compound is therefore written Na_2O_2, and is called sodium *peroxide*: it contains Na^+ cations and peroxide anions, O_2^{2-}.

With potassium, rubidium and caesium, the situation is different again. Whereas lithium and sodium in excess oxygen yield Li_2O and Na_2O_2, potassium, rubidium and caesium form bright orange solids with the formulae KO_2, RbO_2 and CsO_2. Again, the crystal structures show that these contain pairs of oxygens, separated in this case by only 128 pm. The compounds are called *superoxides* and are taken to be assemblies of M^+ and the superoxide ion, O_2^-.

The normal oxides, peroxides and superoxides of the alkali metals all react with water to give alkaline solutions of the hydroxides; for example

$$K_2O(s) + H_2O(l) = 2K^+(aq) + 2OH^-(aq) \qquad (25.2)$$

$$Na_2O_2(s) + 2H_2O(l) = 2Na^+(aq) + 2OH^-(aq) + H_2O_2(aq) \qquad (25.3)$$

$$2KO_2(s) + 2H_2O(l) = 2K^+(aq) + 2OH^-(aq) + H_2O_2(aq) + O_2(g) \qquad (25.4)$$

Notice that peroxides and superoxides give an aqueous solution of hydrogen peroxide, H_2O_2. This decomposes gradually to water and oxygen:

$$H_2O_2(aq) = H_2O(l) + \tfrac{1}{2}O_2(g) \qquad (25.5)$$

How many moles of oxygen gas, $O_2(g)$, are ultimately produced by the reaction of two moles of $KO_2(s)$ with excess water?

1.5 mol; one mole is produced quickly by Equation 25.4, and another 0.5 mol by the slow subsequent decomposition of the mole of H_2O_2 also produced in the reaction.

This 1.5 mol of $O_2(g)$ can also be released by direct reaction of the solid superoxide with carbon dioxide:

$$2KO_2(s) + CO_2(g) = K_2CO_3(s) + \tfrac{3}{2}O_2(g) \qquad (25.6)$$

Figure 25.5
In the Salyut 6 spacecraft, potassium superoxide, KO_2, consumed exhaled CO_2 and replaced it with oxygen.

● What circumstances can you envisage in which this reaction, and therefore KO_2, would be very useful?

● A solid that can convert exhaled $CO_2(g)$ into breathable oxygen could extend the lifetime of a life-support system in a spacecraft or submarine.

Russian spacecraft (Figure 25.5) have, in fact, used superoxides for just this purpose.

25.2.1 The relative stability of peroxides and oxides

Let's look more closely into why it is that when we burn lithium and sodium in air, lithium forms the normal oxide, Li_2O, but sodium yields the peroxide, Na_2O_2. Firstly, we should point out that lithium peroxide, Li_2O_2, does exist; however, it decomposes to the normal oxide on heating to about 200 °C:

$$Li_2O_2(s) = Li_2O(s) + \tfrac{1}{2}O_2(g) \tag{25.7}$$

Sodium peroxide, on the other hand, decomposes in this way only when heated to about 600 °C.

● Use Equation 25.7 to suggest why the products of lithium and sodium combustion are Li_2O and Na_2O_2, respectively.

● The temperatures generated by combustion lie below the decomposition temperature of Na_2O_2, but above that of Li_2O_2.

Thus, the difference in the combustion products arises because the decomposition temperature of Na_2O_2 substantially exceeds that of Li_2O_2. We can now look at this problem thermodynamically. In Section 17.1, you were told that the thermodynamic decomposition temperature of a substance is given, to a good approximation, by

$$T = \frac{\Delta H_m^{\ominus}(298.15\,\text{K})}{\Delta S_m^{\ominus}(298.15\,\text{K})} \tag{17.3}$$

where, in this case, $\Delta H_m^{\ominus}(298.15\,\mathrm{K})$ and $\Delta S_m^{\ominus}(298.15\,\mathrm{K})$ refer to a decomposition reaction of the type

$$M_2O_2(s) = M_2O(s) + \tfrac{1}{2}O_2(g) \tag{25.8}$$

⬤ What can you say about the values of $\Delta S_m^{\ominus}$ for Reaction 25.8 in the two cases (M = Li and M = Na)?

⬤ The two reactions are analogous in the sense defined in Section 16; their $\Delta S_m^{\ominus}$ values will be very similar.

⬤ What factor in Equation 17.3 leads to the decomposition temperature of Na_2O_2 being greater than that of Li_2O_2?

⬤ Because the $\Delta S_m^{\ominus}$ values are similar, the $\Delta H_m^{\ominus}$ value must be larger for the sodium compound; this is so, as the data in Table 25.1 show.

Table 25.1 Values of $\Delta H_m^{\ominus}$ and $\Delta S_m^{\ominus}$ for Reaction 25.8 for M = Li and M = Na

	Li	Na
$\Delta H_m^{\ominus}/\mathrm{kJ\,mol^{-1}}$	36.4	96.7
$\Delta S_m^{\ominus}/\mathrm{J\,K^{-1}\,mol^{-1}}$	83.7	82.7

Thus, if we can explain this difference in $\Delta H_m^{\ominus}$ values, we shall have explained why the decomposition temperature is higher for Na_2O_2 than for Li_2O_2, and why the combustion products are Li_2O for lithium and Na_2O_2 for sodium.

To obtain such an explanation, we use the thermodynamic cycle in Figure 25.6. The decomposition reaction is written along the top; this is the *direct* route from the peroxide to normal oxide plus oxygen; like all reactions in such cycles, the enthalpy change, $\Delta H_m^{\ominus}$, refers to the reaction in the direction of the arrow.

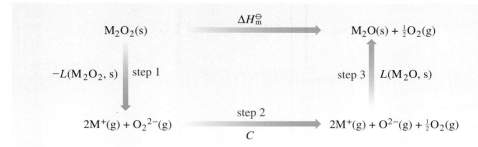

Figure 25.6 Thermodynamic cycle for the thermal decomposition of alkali metal peroxides.

Now let us take an indirect route from peroxide to normal oxide by going round the cycle in an anticlockwise direction. There are three steps:

Step 1 The peroxide, M_2O_2, *breaks down* into its gaseous ions. Notice that this step is the *reverse* of the reaction in which the solid is *formed* from its gaseous ions and for which the standard enthalpy change is the lattice energy $L(M_2O_2, s)$. Thus, the standard enthalpy change for this breakdown reaction is $-L(M_2O_2, s)$.

Step 2 The gaseous ions, $M^+(g)$, appear on both sides of this second step so they can be ignored. The reaction then becomes one in which the gaseous peroxide ion breaks down into a gaseous oxide ion and oxygen gas:

$$O_2^{2-}(g) = O^{2-}(g) + \tfrac{1}{2}O_2(g) \qquad (25.9)$$

The standard enthalpy change is simply labelled C in Figure 25.6.

Step 3 In this final step, the solid oxide $M_2O(s)$ is formed from its gaseous ions; the standard enthalpy change is the lattice energy $L(M_2O, s)$. Application of Hess's law shows that the enthalpy changes by the direct and indirect routes must be equal:

$$\Delta H_m^{\ominus} = -L(M_2O_2, s) + L(M_2O, s) + C \qquad (25.10)$$

Now we use the Kapustinskii equation, Equation 20.11 to obtain values for the lattice energies:

$$L = -\frac{WvZ_+Z_-}{r_+ + r_-} \qquad (20.11)$$

● What are the values of v, Z_+ and Z_- for M_2O_2 and M_2O?

● In both cases, there are three ions in the formula unit, the cation charge is +1 and the anion charge is –2. Thus, $v = 3$, $Z_+ = 1$ and $Z_- = 2$, and

$$\Delta H_m^{\ominus} = \frac{6W}{r(M^+) + r(O_2^{2-})} - \frac{6W}{r(M^+) + r(O^{2-})} + C \qquad (25.11)$$

Now suppose we start with a decomposition reaction in which M is lithium, and we change to one in which M is sodium. Will $\Delta H_m^{\ominus}$ increase or decrease? Clearly, that will depend on how the three terms on the right-hand side of Equation 25.11 change.

● How will the third term, C, be affected by the change from lithium to sodium?

● It will not be affected at all. This is because the metals are not involved in Equation 25.9, for which C is the enthalpy change.

Thus, the changes in $\Delta H_m^{\ominus}$ when sodium is substituted for lithium will depend only on the changes in the first two terms on the right-hand side of Equation 25.11 — that is, on the difference between $6W/[r(M^+) + r(O_2^{2-})]$ and $6W/[r(M^+) + r(O^{2-})]$.

● Will $6W/[r(M^+) + r(O_2^{2-})]$ increase or decrease when M changes from Li to Na?

● It will decrease; Table 23.1 shows that the ionic radius of Na^+, $r(Na^+)$, is greater than that of Li^+, $r(Li^+)$ — 116 pm against 90 pm. This increases the denominator of the fraction where the ionic radii appear.

Likewise, the quantity that is subtracted from this, $6W/[r(M^+) + r(O^{2-})]$ will also decrease. Thus, whether $\Delta H_m^{\ominus}$ increases or decreases when M changes from Li to Na will depend on whether $6W/[r(M^+) + r(O_2^{2-})]$ or $6W/[r(M^+) + r(O^{2-})]$ decreases more. In deciding which, it is crucial to note that, because O_2^{2-} contains two oxygens and O^{2-} contains only one, O_2^{2-} will take up a lot more space in its compounds than does O^{2-}; so $r(O_2^{2-})$ will be considerably greater than $r(O^{2-})$.

● When M changes from Li to Na, which quantity decreases more, $6W/[r(M^+) + r(O_2^{2-})]$ or $6W/[r(M^+) + r(O^{2-})]$?

165

● $6W/[r(M^+) + r(O^{2-})]$; because $r(O^{2-})$ is smaller than $r(O_2^{2-})$, the increase in $r(M^+)$ has the greater impact on this term.

If you find this difficult to follow, look at Figure 25.7, which takes you through an extreme, hypothetical case with real numbers. This also shows that because the larger decrease occurs in $6W/[r(M^+) + r(O^{2-})]$, which has a minus sign in front of it in Equation 25.11, $\Delta H_m^{\ominus}$ becomes more positive when the cation size is increased.

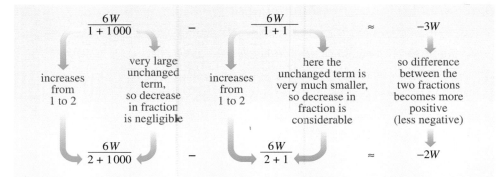

Figure 25.7 A demonstration of the effect that dictates the change in $\Delta H_m^{\ominus}$ for the decomposition of an alkali metal peroxide when the size of the cation is increased. The expression for $\Delta H_m^{\ominus}$ includes a difference between two fractions with identical numerators, $6W$. The denominators of both fractions are the sum of two terms, which we shall call $(A + B)$ for the first fraction and $(A + C)$ for the second. A is common to both terms and $B \gg C$. Illustrating with an extreme case, if $B = 1\,000$ and $C = 1$, then when A is increased from 1 to 2, the decrease in the first fraction is negligible, but that in the second is much larger and results in the total expression becoming more positive.

The principles developed in this Section can be applied in many other situations. What we have shown is that the thermodynamic stabilities and decomposition temperatures of alkali metal peroxides increase with cation size. The argument depended on the fact that the anion in the decomposition product, O^{2-}, was smaller than the anion in the decomposing solid, O_2^{2-}. Because of this, the value of $-L$ for the normal oxide product was greater than that for the decomposing peroxide, and it was also more substantially lowered by an increase in $r(M^+)$.

This suggests a general principle that is so often true that it is worth remembering. Suppose the compound of an alkali metal cation decomposes, through a breakdown of the anion, into a compound of the cation with a new anion. Then if the value of $-L$ for this product exceeds that of the decomposing compound, the stability of the compound to decomposition will increase with cation size.

25.3 Hydrides and nitrides of the alkali metals

When heated quite gently in hydrogen gas, all of the alkali metals form colourless crystalline hydrides, MH. All five compounds have the sodium chloride structure shown in Figure 25.8. When lithium hydride, LiH, is heated, it melts at 680 °C, and if the melt is electrolysed, liquid lithium is formed at one electrode, and hydrogen gas at the other:

negative electrode: $Li^+(melt) + e^- = Li(l)$ (25.12)

positive electrode: $H^-(melt) = \frac{1}{2}H_2(g) + e^-$ (25.13)

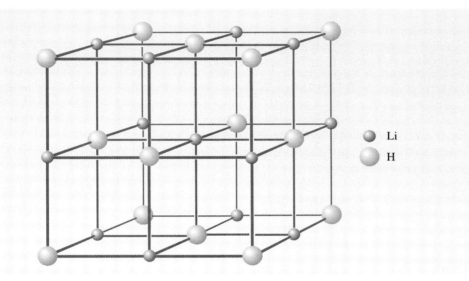

Li

H

Figure 25.8 The alkali metal hydrides all have the sodium chloride structure: in LiH, each lithium ion is octahedrally surrounded by six hydride ions, and each hydride ion is surrounded by six lithium ions.

This behaviour, together with the crystal structure, is characteristic of ionic compounds, and the alkali metal hydrides are usually formulated M^+H^-.

When the alkali metal hydrides are dropped into water, they react to produce hydrogen gas; for example

$LiH(s) + H_2O(l) = Li^+(aq) + OH^-(aq) + H_2(g)$ (25.14)

During the Second World War, lithium hydride was sometimes used to inflate air–sea rescue equipment: Reaction 25.14 occurs on contact with seawater.

All the alkali metal hydrides decompose into hydrogen and the metal when strongly heated:

$MH = M + \frac{1}{2}H_2$ (25.15)

Lithium hydride, however, is by far the most stable with respect to this reaction. It melts at 680 °C, and decomposition begins at around 900–1 000 °C; by contrast, sodium and potassium hydride decompose at about 400 °C. This decrease in stability down the Group appears in an even more marked form when the metals are heated in nitrogen gas. Only lithium reacts, forming a brick-red nitride, Li_3N; the nitrides of other alkali metals are much less stable with respect to the metal and nitrogen.

We shall not undertake a detailed analysis of the thermodynamics of these phenomena in the way that we did for oxygen compounds; instead we shall just summarize the essential conclusions. According to the Born–Haber cycle, when we compare salts of formula type M^+X^-, the factor that tends to stabilize the salt with respect to decomposition *into its constituent elements* is its lattice energy: other things being equal, the more negative the lattice energy (the larger $-L$), the more stable the salt will be. Now, according to Equation 20.11, the value of $-L$ for an alkali metal hydride is $2W/[r(M^+) + r(H^-)]$. The hydride ion, H^- is one of the smallest negative ions, as one would expect of an anion containing only two electrons. Tables of ionic radii, including those in the *Data Book*, show that its ionic radius (126 pm) is similar to that of O^{2-}.

● What does this suggest about the changes in $-L$ for alkali metal hydrides when the cation radius, $r(M^+)$, increases?

● The value of $2W/[r(M^+) - r(H^-)]$ will fall because of the increase in $r(M^+)$, and because $r(H^-)$ is small for an anion, it will fall away particularly steeply.

It is this steep fall in the value of $-L(MH, s)$ that leads to the decrease in the stability of the alkali metal hydrides down Group I. Indeed, one can draw the general conclusion that the stability of alkali metal compounds containing small anions, with respect to their constituent elements, will tend to decrease down the Group.

25.4 Summary of Section 25

1　The alkali metal halides are of formula type MX, and dissolve in water to yield $M^+(aq)$ and $X^-(aq)$ ions. They are colourless crystalline solids, which melt above 450 °C to give conducting melts. CsCl, CsBr and CsI have the eight-coordinate CsCl structure; the rest have the NaCl structure.

2　When the alkali metals burn freely in air, lithium forms a normal oxide, Li_2O. However, sodium yields a peroxide, Na_2O_2, containing O_2^{2-} ions, and potassium, rubidium and caesium form orange superoxides, MO_2, containing O_2^- ions.

3　In water, all three types of oxygen compound form aqueous solutions of hydroxides; the peroxides also yield H_2O_2, and the superoxides form H_2O_2 and oxygen gas. The peroxides and superoxides combine with CO_2, liberating oxygen.

4　Sodium peroxide decomposes to the oxide and oxygen at a higher temperature than Li_2O_2. This can be thermodynamically understood by considering $-L(M_2O_2)$ and $-L(M_2O)$, which are measures of the stabilities of the two types of oxygen compound with respect to their gaseous ions. Because $r(O_2^{2-})$ is greater than $r(O^{2-})$, $-L(M_2O_2)$ is less than $-L(M_2O)$, and decreases by a smaller amount when $r(M^+)$ increases.

5　The preceding result can be generalized: if a solid ionic compound undergoes decomposition by breakdown of the anion into a solid compound with a larger $-L$ value, the stability of the compound tends to increase with cation size.

6　The alkali metals react with hydrogen to form colourless crystalline hydrides, M^+H^-, with the NaCl structure; these react with water to give aqueous hydroxides and hydrogen gas. The stability of the hydrides to decomposition into the constituent elements decreases down the Group: KH and NaH decompose at around 400 °C, but LiH melts at about 700 °C without decomposition, to give a conducting melt that yields lithium and hydrogen at the electrodes during electrolysis.

7   When the metals are heated in nitrogen, only lithium forms a nitride. A major contributor to the decrease in the stabilities of the nitrides and hydrides down the Group is the small size of the anions, H^- and N^{3-}. This makes the decrease in $-L(MH)$ and $-L(M_3N)$ with increasing cation size especially large.

QUESTION 25.1

Pure sodium peroxide is colourless, but the commercial product obtained by burning sodium in a free supply of air is pale yellow. When it is added to ice-cold water, it dissolves, and a small volume of oxygen gas is immediately evolved. This volume is increased some five- to tenfold if the solution is boiled. Suggest an explanation for these observations.

QUESTION 25.2

Lithium hydride is the most stable of the alkali metal hydrides with respect to decomposition into the metal and hydrogen gas, and the decomposition temperatures of the alkali metal hydrides tend to decrease down Group I.

(a)   Explain how this is consistent with the trend in $\Delta H_f^{\ominus}$ (MH, s) values in Table 25.2 for lithium, sodium and caesium.

(b)   How and why does the trend in $\Delta H_f^{\ominus}$ (MI, s) values differ?

Table 25.2 Values of $\Delta H_f^{\ominus}$ (MH, s) and $\Delta H_f^{\ominus}$ (MI, s) for the hydrides and iodides of lithium, sodium and caesium

Alkali metal	$\dfrac{\Delta H_f^{\ominus}(MH,\ s)}{kJ\ mol^{-1}}$	$\dfrac{\Delta H_f^{\ominus}(MI,\ s)}{kJ\ mol^{-1}}$
lithium	−90.5	−270.4
sodium	−56.3	−287.8
caesium	−54.2	−346.6

QUESTION 25.3

It is possible to make compounds of the anion ClF_4^- by heating an alkali metal fluoride with chlorine trifluoride:

$$MF(s) + ClF_3(g) = MClF_4(s) \qquad\qquad (25.16)$$

However, the reaction does not occur with all the alkali metal fluorides. Which fluoride do you think would be most suitable, and why?

METAL IONS, LIGANDS AND COMPLEXES

26

So far, we have denoted metal ions in aqueous solution simply by a parenthesized 'aq' after the formula of the ion, as in $Na^+(aq)$, $Mg^{2+}(aq)$ and $Al^{3+}(aq)$, etc. We have now reached a point where it will be useful to look at the interaction between the metal ion and the surrounding water molecules in a more detailed way. This will lead us to a new concept of great importance in modern inorganic chemistry: the concept of a *metal complex*. Although this Book is concerned with the chemistry of typical elements, such as the alkali metals and alkaline earth metals, it is convenient to introduce this new subject by using a transition element. This is because, in transition-metal chemistry, some of the important changes that we shall be discussing are often marked by changes in colour.

26.1 Metal complexes

Copper sulfate is usually sold as $CuSO_4.5H_2O$. This blue solid (Figure 26.1) is called a **hydrate** because, as the formula implies, there are discrete water molecules distributed throughout the structure. In terms of an ionic model, therefore, the structure contains three types of chemical species: Cu^{2+} ions, SO_4^{2-} ions and water molecules. The arrangement revealed by X-ray diffraction is shown in Figure 26.2. Read the caption carefully.

Figure 26.1
Copper sulfate pentahydrate, $CuSO_4.5H_2O$ (left), and its blue solution in water (right).

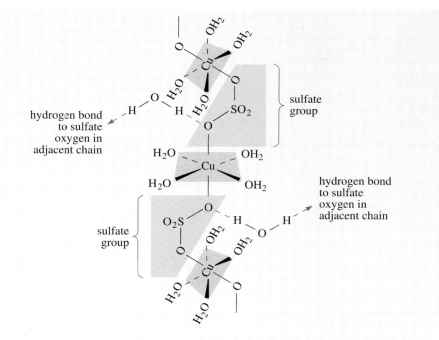

Figure 26.2 Structural features of $CuSO_4.5H_2O$. Four of the five water molecules in each formula unit are arranged around the copper ion at the corners of a square, which is picked out in blue. If we take the copper to be a +2 ion, the formula of the square unit is $Cu(H_2O)_4^{2+}$. Above and below the squares are oxygen atoms of sulfate ions through which the $Cu(H_2O)_4^{2+}$ units are linked into chains. The fifth water molecule in each formula unit is not bound to copper: it binds adjacent chains together by means of the hydrogen bonds that it forms with sulfate oxygens.

The structure can be broken down into chains made up of alternate $Cu(H_2O)_4^{2+}$ and SO_4^{2-} ions. The chains are held together by hydrogen bonds formed through one of the five water molecules in each formula unit. Let's try to justify this breakdown further.

⬤ In Figure 26.2, lines emanating from the copper ion show its bonding with surrounding atoms. How many, and of what type are these surrounding atoms?

⬤ Six lines emanate from each copper ion and each one links it to an oxygen atom. Four of the six surrounding oxygen atoms belong to water molecules; the other two belong to sulfate ions.

Figure 26.2 also tries to show that the arrangement of the six oxygens around the copper is octahedral. To clarify this, the octahedron is shown enlarged in Figure 26.3, with the copper–oxygen bond lengths included.

⬤ Figure 26.3 suggests that some of the oxygens are more closely and tightly bound to the Cu^{2+} ion than others. How many of these closely bound oxygens are there, what is their type, and how are they arranged around the copper?

⬤ The four oxygens of the four water molecules are 40 pm closer to the Cu^{2+} ion than the two sulfate oxygens; the four are arranged at the corners of a square, which is picked out in blue in Figure 26.2.

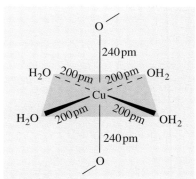

Figure 26.3
The Cu–O distances in the octahedron of oxygen atoms around the copper in $CuSO_4.5H_2O$. Four of these oxygens belong to water molecules; the other two to sulfate ions.

171

This especially close binding between Cu^{2+} and four water molecules is further justification for regarding them as a separate unit. It suggests a new way of looking at $CuSO_4.5H_2O$: it contains ions with the formula $[Cu(H_2O)_4]^{2+}$, in which a Cu^{2+} ion is bound to four waters at the corners of a square in *square-planar coordination* as in Structure **26.1**. Thus, instead of saying that the solid consists of one Cu^{2+} ion and five water molecules for every SO_4^{2-} ion, we say that it contains one $[Cu(H_2O)_4]^{2+}$ ion, and one water molecule for every sulfate ion. The entity $[Cu(H_2O)_4]^{2+}$ is known as a **complex**, or, because it carries a charge in this case, as a *complex ion*. Note the use of square brackets enclosing the formula of the complex, the charge being placed outside them.

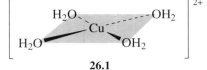

26.1

Some confirmation that this is a sensible way of looking at $CuSO_4.5H_2O$ comes when we dissolve the solid in water. The blue solid gives a blue solution of similar hue (Figure 26.1). We can explain the common blue coloration by arguing that it is a property of the complex ion $[Cu(H_2O)_4]^{2+}$, which persists when the solid dissolves in aqueous solution. Up till now, we would have written the dissolution of $CuSO_4.5H_2O$ in water as:

$$CuSO_4.5H_2O(s) = Cu^{2+}(aq) + SO_4^{2-}(aq) + 5H_2O(l) \qquad (26.1)$$

○ Write an alternative equation for this change incorporating our explanation of the similar colours of $CuSO_4.5H_2O(s)$ and its aqueous solutions.

○ The equation is:

$$[Cu(H_2O)_4]SO_4.H_2O(s) = [Cu(H_2O)_4]^{2+}(aq) + SO_4^{2-}(aq) + H_2O(l) \qquad (26.2)$$

Notice that we still write '(aq)' after the formula of the complex ion in aqueous solution. Although four water molecules are bound especially closely by the Cu^{2+} ion, the resulting complex, $[Cu(H_2O)_4]^{2+}$, still interacts with surrounding solvent water molecules that are further away from the central copper.

Another sign of the value of these ideas is revealed when an aqueous solution of ammonia, $NH_3(aq)$, is added to our solution of copper sulfate. The light blue solution turns an intense deep violet (Figure 26.4), showing that some new complex has been formed. What happens is that each water molecule at the corner of the square in Structure **26.1** is replaced by an ammonia molecule:

$$\begin{bmatrix} & H_2O & \\ H_2O- & \!\!Cu\!\! & -OH_2 \\ & H_2O & \end{bmatrix}^{2+} (aq) + 4NH_3(aq) = \begin{bmatrix} & NH_3 & \\ H_3N- & \!\!Cu\!\! & -NH_3 \\ & NH_3 & \end{bmatrix}^{2+} (aq) + 4H_2O(l) \qquad (26.3)$$

light blue   deep violet

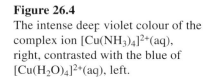

Figure 26.4
The intense deep violet colour of the complex ion $[Cu(NH_3)_4]^{2+}(aq)$, right, contrasted with the blue of $[Cu(H_2O)_4]^{2+}(aq)$, left.

By recognizing the existence of complexes, we obtain a simple but elegant description of the reaction: the copper ion retains its preference for square-planar coordination, but changes the four atoms to which it is bound from oxygen to nitrogen, by exchanging water for ammonia.

Ammonia is not the only molecule that can effect a substitution of nitrogen for oxygen around Cu^{2+}. An especially interesting case occurs when an aqueous solution of the compound commonly called ethylenediamine (Structure **26.2**) is added to an aqueous solution of copper sulfate. This molecule contains *two* nitrogen atoms, and its shape and size are such that they can comfortably occupy two adjacent corners of the square-planar coordination around a Cu^{2+} ion in a complex.

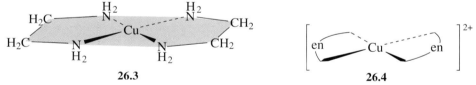

$$H_2N \quad\quad NH_2$$
$$\backslash \quad\quad\quad /$$
$$CH_2 - CH_2$$

26.2

○ Assuming that Cu^{2+} retains its preference for square-planar coordination, and that ethylenediamine reacts with $[Cu(H_2O)_4]^{2+}$ to form a new complex, how many ethylenediamine molecules will react with each $[Cu(H_2O)_4]^{2+}$?

○ Two; there are four positions to be occupied around the copper ion, and each ethylenediamine takes care of two of them:

$$\left[\begin{array}{c} H_2O \\ | \\ H_2O - Cu - OH_2 \\ | \\ H_2O \end{array}\right]^{2+} (aq) \;+\; 2 \begin{array}{c} CH_2NH_2 \\ | \\ CH_2NH_2 \end{array} \;=\; \left[\begin{array}{c} H_2C \\ H_2C \quad\; NH_2 \\ \backslash\quad\quad | \\ H_2N - Cu - NH_2 \\ | \\ H_2N \quad\; CH_2 \\ \quad CH_2 \end{array}\right]^{2+} (aq) \;+\; 4H_2O(l) \quad\quad (26.4)$$

The complex that is formed (Structure **26.3**) again has a deep blue colour. Ethylenediamine is often abbreviated to 'en'; the complex can then be drawn as in Structure **26.4**.

$$\begin{array}{cc}
H_2C - N(H_2) - Cu - N(H_2) - CH_2 \\
H_2C - N(H_2) \quad\quad N(H_2) - CH_2
\end{array}$$

26.3

$$\left[\; en \;\diagup\; Cu \;\diagdown\; en \;\right]^{2+}$$

26.4

26.2 The anatomy of a complex

In this Section we look more closely at the components of a complex, and the way in which they are bound together. In the cases we shall consider, there is a central atom or ion such as Cu^{2+}, and this is bound to and surrounded by groups such as H_2O, NH_3 or en. These groups are called **ligands**. Each ligand has one or more **coordinating atoms**, which are the atoms that are bound to the central metal ion. In the case of water this is oxygen; in the case of ammonia, it is nitrogen. Ligands may be neutral molecules or ions. For example, the halide ions, F^-, Cl^-, Br^- and I^-, act as ligands in many metal complexes.

○ If concentrated hydrochloric acid is added to an aqueous solution containing copper ions, the four water molecules in $[Cu(H_2O)_4]^{2+}$ are replaced by chloride ligands. Write an equation for the reaction, making sure that the charge is balanced.

○ The equation is:

$$[Cu(H_2O)_4]^{2+}(aq) + 4Cl^-(aq) = [CuCl_4]^{2-}(aq) + 4H_2O(l) \quad\quad (26.5)$$

Note how the substitution of charged ligands for neutral ones alters the overall charge of the complex: from an ionic standpoint, $[CuCl_4]^{2-}$ is an assembly of one Cu^{2+} and four Cl^- ions, making the overall charge -2.

Now let us turn to the bonding between the metal ion and the ligands. The ligands that we have introduced are in states in which all the atoms have noble gas configurations. This is clear from Figure 26.5, which shows Lewis structures for three typical examples: H_2O, NH_3 and F^-. Notice first that the coordinating atoms in the ligands are all electronegative (O, N and F).

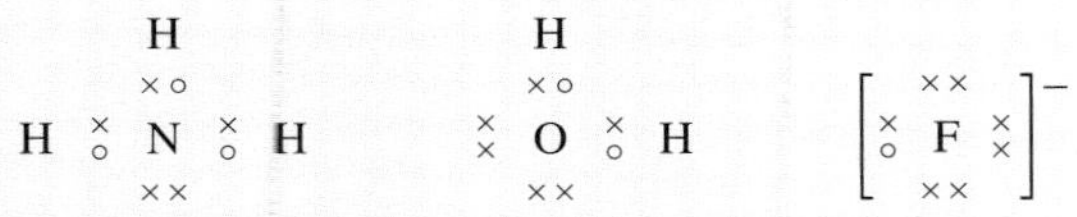

Figure 26.5 Lewis structures for the ligands NH_3, H_2O and F^-; by convention, the electrons, whether bonding or non-bonding, are grouped in pairs.

 In Figure 26.5, what else do the coordinating atoms have in common?

 They all carry non-bonded electron pairs that are are not involved in chemical bonding. For example, nitrogen in NH_3 has one non-bonded pair, and oxygen in H_2O has two.

These observations suggest *two* crude but useful ways of looking at the bonding between the metal atom, or ion, and the ligand. Firstly, there is an electrostatic viewpoint: the central metal atom carries a net positive charge, so the ligands orientate themselves with the negative charge of the non-bonded pairs directed towards this positive site; the electrical interaction between them binds the metallic element and the ligand together. Secondly, there is a covalent viewpoint: the ligands' non-bonded pairs become electron pair bonds if the ligands donate them to the metallic element.

Ligands such as H_2O, NH_3 and F^-, which have just one coordinating atom — one point of attachment through which they can bind to a metal atom or ion — are called **unidentate ligands**.

 What do you think a ligand like ethylenediamine is called?

 It contains *two* nitrogen atoms with non-bonded electron pairs — two coordinating atoms — and is called a **bidentate ligand**.

There are also **polydentate ligands**, which contain *more than two* coordinating atoms. Table 26.1 includes diethylenetriamine (dien), which is tridentate, and the hexadentate ethylenediaminetetraacetate anion ($edta^{4-}$). In this latter case, the coordinating atoms are the two nitrogens, and four negatively charged oxygen atoms of the four $-CH_2COO^-$ groups. The geometry of $edta^{4-}$ is such that these six atoms, with their non-bonded pairs of electrons, can occupy all six positions around a metallic element in an octahedral complex (see Figure 28.19).

Table 26.1 A few bidentate and polydentate ligands

Name	Abbreviation	Structure
ethylenediamine	en	H_2N \ NH_2 / CH_2-CH_2
oxalate	ox^{2-}	O O $C-C$ ^-O O^-
diethylenetriamine	dien	H_2N \ $\overset{H}{N}$ / NH_2 $(CH_2)_2$ $(CH_2)_2$
ethylenediaminetetraacetate	$edta^{4-}$	$H_2C-N(CH_2COO^-)_2$ | $H_2C-N(CH_2COO^-)_2$

* Notice that chemists who work with metal complexes often use names and abbreviations for ligands that pre-date modern, systematic organic nomenclature; for example, ethylenediamine (en) has the systematic name 1,2-diaminoethane.

26.3 Summary of Section 26

1 Metallic elements are often found in a combined state as complexes. A complex usually consists of a central metal atom or ion bound to surrounding molecules or ions called ligands. Common ligands include halide ions, water, and ammonia.

2 Ligands are linked to the metal through coordinating atoms, such as oxygen, nitrogen or halogen, which are electronegative and carry non-bonded electron pairs.

3 A ligand may contain more than one coordinating atom. For example, en, dien and $edta^{4-}$ contain two, three and six coordinating atoms, respectively, and are said to be bidentate, tridentate and hexadentate, respectively.

QUESTION 26.1

Magnesium sulfate occurs as a colourless hydrate, $MgSO_4.7H_2O$, often called Epsom salts. Part of the structure is shown in Figure 26.6. Study the key carefully, and then answer the following questions.

(a) The distances between the magnesium ion and its immediate neighbours are nearly identical. Write the formula of the magnesium complex that this solid contains. How do you think the ligands are arranged around the magnesium ion?   OCTAHEDRALLY DISPOSED AROUND Mg

(b) Assuming that the complex persists in the aqueous solution when $MgSO_4.7H_2O$ dissolves in water, write an equation for the dissolving process.

(c) A new complex is formed when this aqueous solution is treated with an aqueous solution of $edta^{4-}$. Use Table 26.1, and your answer to part (a) of this question, to write an equation for the reaction that occurs.

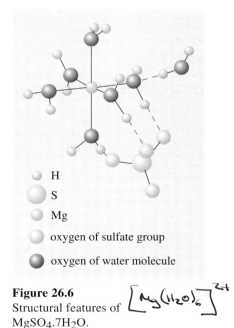

H
S
Mg
oxygen of sulfate group
oxygen of water molecule

Figure 26.6 Structural features of $[Mg(H_2O)_6]^{2+}$ $MgSO_4.7H_2O$.

175

ALKALI METAL COMPLEXES

27

Until quite recently, alkali metals were not thought of as elements whose cations formed clearly defined complexes that would persist in a sequence of chemical reactions. In particular, it was not easy to find evidence for alkali metal complexes, either in solution or in the solid state, in which the cation was bound to ligands that were not shared with other cations. For example, when sodium sulfate is crystallized from water, it does so as the hydrate $Na_2SO_4.10H_2O$. In this hydrate, each sodium ion is surrounded by six water molecules at the corners of an octahedron, but one cannot describe this unit as a $[Na(H_2O)_6]^+$ complex because four of the six water molecules are shared with, and bound to, other sodium ions.

This situation changed during the late 1960s with the discovery by Charles Pedersen (Figure 27.1) of ligands known as **crown ethers**. Typical examples of crown ethers are shown as Structures **27.1** and **27.2**.

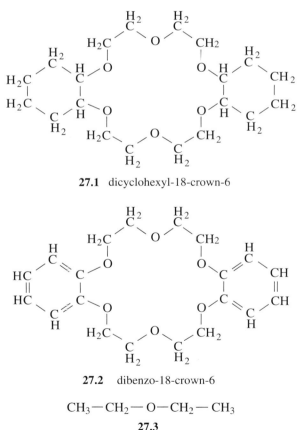

27.1 dicyclohexyl-18-crown-6

27.2 dibenzo-18-crown-6

$$CH_3-CH_2-O-CH_2-CH_3$$
27.3

- Which sites in Structure **27.1** are likely to enable it to act as a ligand, and how many of them are there?

- The six oxygen atoms carry non-bonded electron pairs like those on the oxygen atom in a water molecule. The molecule is potentially a hexadentate ligand.

Figure 27.1
Charles Pedersen (1904–1989), the son of a Norwegian seaman and mining engineer, was born at Pusan, Korea and went to school in Japan. At the age of 17, he travelled alone to the United States where he took degrees at the University of Dayton, Ohio and the Massachusetts Institute of Technology. In 1927, he joined DuPont, and in the 34th of the 42 years he spent with this company, he obtained some unexpected white needle-like crystals while working on organic catalysts. Instead of throwing them away, he studied their properties, and found that in their presence, sodium hydroxide would dissolve in organic solvents such as diethyl ether (Structure **27.3**) or benzene (Structure **27.4**). The crystals turned out to be those of the crown ether **27.2**, which he named dibenzo-18-crown-6 because the central ring bridges two benzene rings and includes eighteen atoms, of which six are oxygen. For the discovery of such crown ethers and their properties, Pedersen shared the 1987 Nobel Prize for Chemistry.

Pedersen's discovery of crown ethers was itself accidental, but it was a second accidental observation that ultimately led to the award of the Nobel Prize for Chemistry. As you know, it is an accepted property of ionic compounds, such as alkali metal salts, that they are almost insoluble in non-polar organic solvents such as heptane, trichloromethane (chloroform) and benzene. In other words, if we try to dissolve potassium iodide in chloroform:

$$KI(s) \rightleftharpoons K^+(org) + I^-(org) \qquad (27.1)$$

we find that the equilibrium lies well over to the left. Pedersen found that, in many cases, this situation was reversed when a crown ether was added to the organic solvent. For example, in the case that we have just discussed, in the presence of the molecule shown as Structure **27.1**, potassium iodide will dissolve in chloroform to the tune of 7 grams per litre.

● How can you explain this by using Le Chatelier's principle?

● If the cyclic ether acts as a ligand and complexes with the alkali metal cation, it will disturb the equilibrium in Equation 27.1, shift it to the right, and increase the solubility of KI in chloroform.

This capacity of crown ethers to make alkali metal salts soluble in organic solvents has proved extremely useful in organic chemistry. For example, potassium permanganate ($KMnO_4$), contains the permanganate ion, MnO_4^-, which will oxidize many organic compounds. However, there are difficulties in bringing the reactants into intimate contact. Solid $KMnO_4$ is likely to react slowly. In trying to bring the reactants together in a common solvent, one finds that organic solvents that dissolve the organic reactant will not dissolve $KMnO_4$, and a solvent like water will dissolve $KMnO_4$, but may not dissolve the organic reactant, or may sometimes even react with it. These difficulties can be circumvented by using benzene (**27.4**) as the organic solvent, and adding both $KMnO_4$ and the cyclic ether **27.1**. The solubility of $KMnO_4$ in benzene is then as much as $30\,g\,litre^{-1}$. Many organic compounds can be dissolved in this benzene solution, and oxidized smoothly and easily at room temperature.

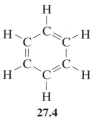

27.4

Solid compounds containing alkali metal complexes with cyclic ether ligands can be crystallized from a solution of an alkali metal salt and a cyclic ether in an organic solvent. For example, Figure 27.2 shows part of the crystal structure of a compound obtained from a solution of sodium thiocyanate, NaNCS, and the crown ether **27.2** in methanol. The sodium ion sits just below, but close to the centre of the inner ring of six oxygens on the crown ether ligand. As expected then, the ligand is hexadentate, and we can think of the complex as $[NaL]^+$, where L is Structure **27.2**. The other close neighbour of the sodium is the nitrogen of the thiocyanate ion, NCS^-, which is some 50 pm further away than the oxygens of the ether. The compound is often said to contain ion pairs, $[NaL]^+[NCS]^-$, in which the cation is the alkali metal complex.

A measure of the stability of such alkali metal complexes in solution is the equilibrium constant of the reaction

$$M^+(solv) + L(solv) = [ML]^+(solv) \qquad (27.2)$$

where 'solv' denotes some general solvent. When L is Structure **27.1**, and the solvent is methanol, the potassium complex (M = K) is the most stable as

Figure 27.3 shows. Now, Figure 27.2 suggests that the alkali metal cation in these complexes sits close to the centre of the encircling ligand. Perhaps then, the potassium complex is the most stable because the potassium ion fits most snugly into the cavity in the cyclic ether. One way of testing this hypothesis is to decrease the size of the cavity by using ligand **27.5**, whose inner ring contains two fewer carbon atoms and two fewer oxygen atoms.

When this change is made, potassium no longer forms the most stable complex. Would you expect sodium or rubidium to form the more-stable complex?

Sodium; the cavity becomes smaller when the ligand change is made. As Na^+ is smaller than K^+ or Rb^+, it is this ion that will now fit more snugly into it.

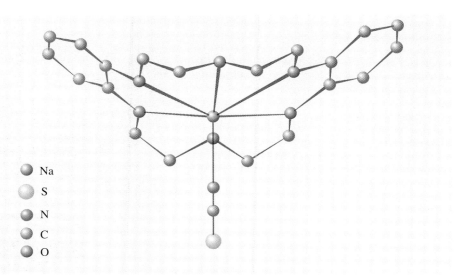

Na

S

N

C

O

Figure 27.2 Structural features of the compound $[NaL]^+ NCS^-$, where L is dibenzo-18-crown-6 (**27.2**); the hydrogen atoms have been omitted for clarity. Sodium is coordinated by the six oxygen atoms of the crown ether, the Na—O distances being about 280 pm. The distance between sodium and the nitrogen atoms in the thiocyanate ion, $^-N{=}C{=}S$, is 332 pm. The diagram shows why such ligands are called crown ethers: the ligand has a crown-like shape and the sodium is topped by the cavity of the crown.

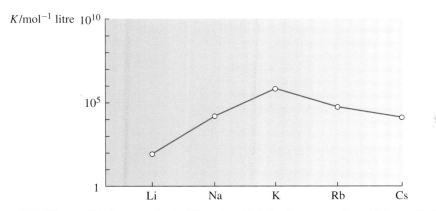

K/mol^{-1} litre

Figure 27.3 The equilibrium constant of Reaction 27.2 for the cases where M is an alkali metal, the solvent is methanol, and the ligand L is dicyclohexyl-18-crown-6 (**27.1**). Note that the scale on the vertical axis is labelled in powers of 10.

178

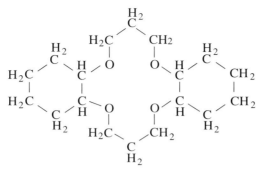

27.5   dicyclohexyl-14-crown-4

Notice, however, that the 'cavity' in crown ethers, such as **27.1**, **27.2** and **27.5**, is a cavity in a ring and is therefore close to two dimensional. Soon after Pedersen published his work on crown ethers, the French chemist, Jean-Marie Lehn (Figure 27.4), set out to modify the cyclic ether structures so that the cavity became three dimensional. The best known of these new ligands is shown as Structure **27.6**.

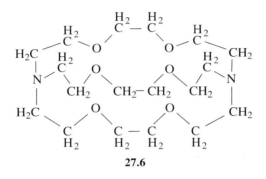

27.6

🔵 How many coordinating atoms does Structure **27.6** contain?

🔵 There are eight: the six oxygens and two nitrogens all carry non-bonding electron pairs.

This ligand contains *three* rings with potential bonding sites, and so the cavity at its centre is three dimensional. The ligand and its bonding atoms can wrap themselves around a cation of the appropriate size, forming a 'crypt', which conceals and protects the cation from surrounding molecules or ions. For this reason, such ligands are known as **cryptands**; Structure **27.6** is called cryptand-222 and written C_{222}. The three twos refer to the three pairs of oxygen atoms which lie on the three chains that link the nitrogen atoms.

One effect of this new arrangement is that the complexes become much more stable. For example, the equilibrium constant of the reaction

$$K^+(\text{methanol}) + C_{222}(\text{methanol}) = [KC_{222}]^+(\text{methanol}) \qquad (27.3)$$

is about $10^{10}\,\text{mol}^{-1}$ litre, about 10 000 times the value with crown ether **27.3** plotted in Figure 27.3. In Figure 27.5, we plot the equilibrium constants for reactions like Equilibrium 27.3 for each of the alkali metal cations; as in Figure 27.3, there is a maximum at potassium. However, sodium ions also form very stable complexes with C_{222}, and, as you will now see, this led to a striking discovery in alkali metal chemistry.

Figure 27.4
Jean-Marie Lehn, Professor of Organic Chemistry at the University of Strasbourg. He developed ligands such as cryptands, which contain three-dimensional cavities. With Charles Pedersen (Figure 27.1), he shared the 1987 Nobel Prize for Chemistry for contributions to the idea of *molecular recognition*. Figure 27.5 provides a very simple example of this idea. Cryptand-222 (**27.6**) forms a complex with potassium whose equilibrium constant (Equation 27.2) is some 100 times greater than that with any other alkali metal cation. The ligand's size and shape are such that it will 'recognize' and preferentially combine with one of a number of otherwise quite similar species.

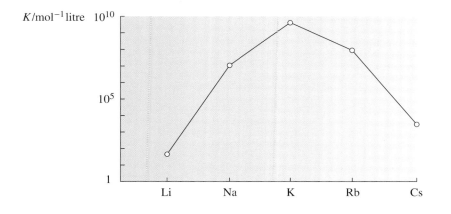

Figure 27.5
The equilibrium constants of Reaction 27.2 for the cases where M is an alkali metal, the solvent is methanol, and the ligand is cryptand-222 (**27.6**).

27.1 Alkali metal anions

In Section 23.1.2, you saw that all five alkali metals dissolve in liquid ammonia to give blue solutions containing ammoniated electrons:

$$M(s) = M^+(amm) + e^-(amm) \qquad (27.4)$$

Since the blue colour, and the absorption spectrum that goes with it (Figure 27.6), is a property of the ammoniated electrons and not of the alkali metals, it is common to solutions of all five alkali metal–ammonia complexes.

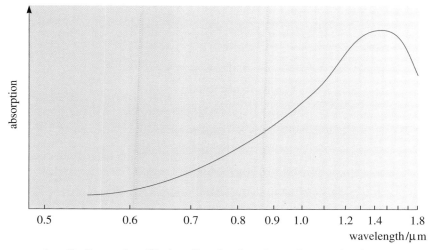

Figure 27.6
The absorption spectrum of the blue solutions obtained by dissolving small amounts of alkali metal in liquid ammonia. The absorption is attributed to the ammoniated electron, $e^-(amm)$.

However, the alkali metals will also dissolve in other solvents that have some resemblance to liquid ammonia. One of these is aminoethane, $C_2H_5NH_2$, which is a liquid at room temperature. Again the solutions have a blue colour, but the spectra for the different metals are quite different: they are shown in Figure 27.7.

The absorption peak for solvated electrons in ammonia and related solvents usually occurs in the wavelength range marked by the pink band in Figure 27.7.

● Which metal most clearly gives solvated electrons?

● Only lithium gives an absorption peak in this region.

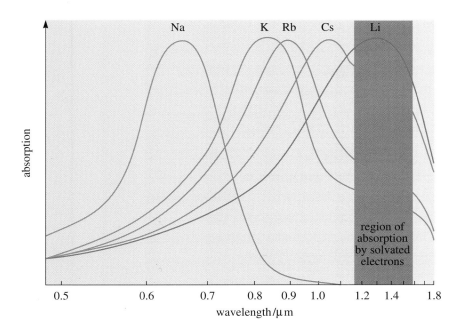

Figure 27.7
The absorption spectra of the solutions obtained when the alkali metals dissolve in aminoethane.

Because sodium, potassium, rubidium and caesium give absorption bands with peaks at distinctly different wavelengths, these bands must be generated by species that contain the different metallic elements.

● Look at Figure 27.6: is it likely that these bands are a property of the solvated cations, M^+?

● Not very likely; no such bands appear in the spectra of alkali metal solutions in liquid ammonia, which contain $M^+(amm)$: the cations appear not to absorb at these wavelengths.

This and other evidence led to the suggestion that, in aminoethane (ameth), the dissolution of the alkali metals takes place chiefly by the route

$$2M(s) \rightleftharpoons M^+(ameth) + M^-(ameth) \tag{27.5}$$

The peaks in the absorption spectra of Na, K, Rb and Cs in aminoethane solution (Figure 27.7) can then be attributed to solvated *alkali metal anions*. This is not quite as outlandish as it might appear: in Table 23.1, we quoted the electron affinities of the alkali metals, and they were positive. This shows that the uptake of electrons by gaseous alkali metal atoms is exothermic; for example

$$Na(g) + e^-(g) = Na^-(g); \quad \Delta H_m^{\ominus} = -53 \, kJ \, mol^{-1} \tag{27.6}$$

The *sign* of the energy change for this step at least, is favourable to the formation of compounds or solutions containing such anions. This is not the case for the noble gases or the Group II elements, which precede and follow, respectively, the alkali metals in the Periodic Table. A possible reason for this is provided by the change in electronic configuration that occurs during a reaction like Equation 27.6.

● What is the electronic configuration of the alkali metal anion Na^-?

● $1s^2 2s^2 2p^6 3s^2$: in Reaction 27.6 an electron is added to the outer ns^1 configuration of the metal atom to form a full ns^2 sub-shell.

In one sense then, the acquisition of an electron by a sodium atom when forming Na^- resembles the acquisition of an electron by a chlorine atom when forming Cl^-: in both cases, a sub-shell is filled (3s in the case of sodium; 3p in the case of chlorine). Of course, in the chlorine case, the filling of the sub-shell completes the stable noble-gas configuration, a fact reflected in the much larger electron affinity ($349\,kJ\,mol^{-1}$). Nevertheless, the analogy, combined with the still fairly favourable $\Delta H_m^{\ominus}$ of Reaction 27.6, makes the idea of compounds containing alkali metal anions more plausible than it might first seem.

How can we prepare such a compound? You have seen that when sodium is added to aminoethane until no more will dissolve, the following equilibrium is set up:

$$2Na(s) \rightleftharpoons Na^+(ameth) + Na^-(ameth) \qquad (27.7)$$

Unfortunately, very little sodium will dissolve in aminoethane: the equilibrium lies well over to the left. One sign of this is that if we try to make compounds of the type Na^+Na^- or K^+K^- by evaporation of solutions of alkali metals in aminoethane, we obtain the original metals rather than compounds containing anions.

How might we shift the equilibrium in Equation 27.7 to the right, and so stabilize the solution containing Na^-?

A crown ether or cryptand that complexes with Na^+ might do the trick.

In the late 1970s, Professor James Dye of Michigan State University used C_{222} for this purpose. In the presence of this ligand, sodium dissolved in aminoethane to form $[NaC_{222}]^+$, and the solubility of the metal was increased to around $5\,g\,litre^{-1}$. The equilibrium can now be written:

$$2Na(s) + C_{222}(ameth) \rightleftharpoons [NaC_{222}]^+(ameth) + Na^-(ameth) \qquad (27.8)$$

When the solution was cooled to $-15\,°C$, shiny golden crystals of the compound $[NaC_{222}]^+Na^-$ were precipitated (Figure 27.8). The compound did not decompose below $0\,°C$ provided it was rigorously protected from air and moisture by keeping it in a dry atmosphere of pure nitrogen or argon. Above $0\,°C$, it tended to decompose to sodium metal:

$$[NaC_{222}]^+Na^- = 2Na(s) + C_{222}(s) \qquad (27.9)$$

Figure 27.8
Gold-coloured crystals of the compound formulated $[NaC_{222}]^+Na^-$, which contains an Na^- ion.

The crystal structure of the complex is shown in Figure 27.9. As you can see, the sodium in the complex cation is surrounded by the C_{222} ligand and bonded to all eight of its donor atoms. Although it is quite difficult to spot, the Na^- ions are arranged around the complex cation at the corners of an octahedron: in the Figure, the Na^- ions have been connected by triangles that form the faces of the octahedron.

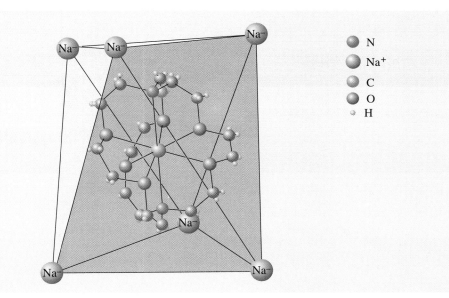

Figure 27.9
The structure of $[NaC_{222}]^+Na^-$. The sodium at the centre lies within the cavity of the cryptand-222, where it is coordinated by six oxygen and two nitrogen atoms. Around the $[NaC_{222}]^+$ complex, there are six Na^- ions at the corners of an octahedron.

Key:
- N
- Na^+
- C
- O
- H

27.2 Summary of Section 27

1 The alkali metal cations form relatively few discrete complexes in solution. However, during the 1960s and 1970s, they were found to form complex ions with polycyclic ethers such as dicyclohexyl-18-crown-6 (**27.1**) and cryptand-222 (**27.6**).

2 If a crown ether is added to benzene, the benzene then dissolves some common alkali metal salts, such as $KMnO_4$, because the complexing between the alkali metal cation and the crown ether makes the otherwise unfavourable dissolution reaction favourable. The resulting solution of MnO_4^- in benzene can be used as an oxidizing reagent in organic chemistry.

3 Such complexes have their greatest stability when the size of the alkali metal cation and the size of the cavity in the ligand are suitably matched.

4 In liquid ammonia the alkali metals form solvated cations, M^+, and solvated electrons; in aminoethane, the major products with sodium, potassium, rubidium and caesium are solvated cations, M^+, and solvated anions, M^-. However, the solubility of the metals in this latter case is very small.

5 The addition of cryptand-222 to the aminoethane and the resulting complexing with the cations, M^+, greatly increases the solubility of the alkali metal. At $-15\,°C$, the sodium solution then deposits crystals of the salt $[NaC_{222}]^+Na^-$, which contains a sodium anion.

QUESTION 27.1

In cryptand-222 (**27.6**), the two nitrogen atoms are linked by three chains of structure $-CH_2-CH_2-O-CH_2-CH_2-O-CH_2-CH_2-$. In cryptand-221, one of these three chains is shortened to $-CH_2-CH_2-O-CH_2-CH_2-$, and in cryptand-211, two of the three chains are so shortened. With cryptand-222 in methanol, it is potassium that forms the most stable alkali metal complex. With cryptand-221, it is a different alkali metal cation, and with cryptand-211, a different cation again. Identify the alkali metal cations that form the most stable complexes in methanol with cryptand-221 and with cryptand-211.

THE GROUP II OR ALKALINE EARTH ELEMENTS

In the final Section of the main text, we discuss the chemistry of the elements beryllium, magnesium, calcium, strontium, barium and radium. They occur in Group II of the Periodic Table (Figure 28.1), and are often called the alkaline earth elements. The word 'earth' is an old-fashioned name for metal oxides of very high melting temperature, which can be obtained by subjecting minerals containing the elements to heat or some other simple process. When the oxides MgO, CaO, SrO, BaO and RaO (but not BeO) are added to water, they form hydroxides that are soluble enough to give solutions that are alkaline (hence the term *alkaline earth element*):

$$MO(s) + H_2O(l) = M(OH)_2(s) \qquad (28.1)$$

$$M(OH)_2(s) = M^{2+}(aq) + 2OH^-(aq) \qquad (28.2)$$

Of the six elements, calcium and magnesium are by far the most common, being the fifth and sixth most abundant elements in the Earth's crust. Both elements occur as dolomite, $CaMg(CO_3)_2$ (Figure 28.2), which is a major constituent of certain mountain ranges, such as the Dolomite Alps in Italy. Magnesium occurs by itself in deposits of magnesite, $MgCO_3$, and after sodium, it is the most abundant metallic element in seawater, of which it constitutes about 0.13% by mass.

The most important sources of calcium are the huge sedimentary deposits of calcium carbonate, $CaCO_3$, which are formed from the fossilized remains of long-dead shellfish. In this category come the minerals limestone, chalk and marble. Dissolved carbon dioxide makes rainwater slightly acid; consequently, limestone is slightly soluble in rain:

$$CaCO_3(s) + H^+(aq) = Ca^{2+}(aq) + HCO_3^-(aq) \qquad (28.3)$$

The resulting erosion gives rise, even at moderate heights, to rugged landscapes of a quality that in other rocks are confined to coastlines and mountains (Figure 28.3). Hard limestone country is often characterized by inland cliffs, caves, gorges and springs. Cheddar Gorge in Somerset and Ingleborough in Yorkshire are fine British examples.

The erosion arising from Reaction 28.3 also develops open pores, cracks and fissures in limestone strata so that they become excellent aquifers (media for the storage and transmission of water). In Britain, limestones are the source of more drinking water than all other aquifers put together. Figure 28.4 shows how limestone strata warm up water from the Mendip Hills as they carry it some 15 km to Bath, where it emerges in the famous hot springs.

The most important beryllium mineral is the aluminosilicate, beryl, $Be_3Al_2(SiO_3)_6$. Emeralds have the same composition, except that they contain about 2% chromium, which provides the green colour. Both strontium and barium occur naturally as the sulfates celestite, $SrSO_4$, and barite, $BaSO_4$. Other Group II minerals are the carbonates strontianite, $SrCO_3$, and witherite, $BaCO_3$.

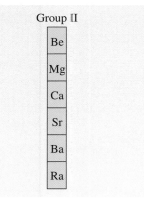

Figure 28.1
The Group II elements.

Figure 28.2
Large crystals of the mineral dolomite.

Figure 28.3
Hard limestone country, such as Cheddar Gorge, is among the most attractive in the British Isles.

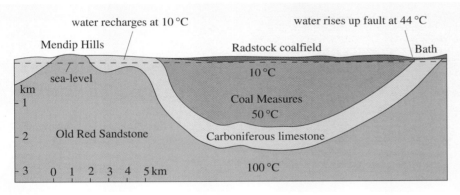

Figure 28.4
Limestone strata transmit water from the Mendip Hills to Bath, warming it up as they do so.

Radium occurs in association with uranium because $^{226}_{88}Ra$, its most long-lived isotope, is one of the intermediate products in the long chain of radioactive decays that converts $^{238}_{92}U$ into $^{206}_{82}Pb$. The isotope $^{226}_{88}Ra$ undergoes α-decay:

$$^{226}_{88}Ra = {}^{222}_{86}Rn + {}^{4}_{2}He \; ; \; t_{1/2} = 1600 \text{ years} \qquad (28.4)$$

About 10 tonnes of uranium ore will yield 1 mg of radium.

28.1 The alkaline earth metals

The redox potentials of the Group II metals, $E^{\ominus}(M^{2+}|M)$, are comparable with those of the alkali metals (Table 21.2). This shows that they are very powerful reducing agents.

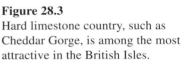 What method is often used to extract such metals from their ores or compounds?

● The electrolysis of molten compounds (Section 17.3).

185

All of the metals are extracted by the electrolysis of their molten chlorides, although other compounds such as potassium or sodium chloride are usually added to the melt to lower the melting temperature and improve the conductivity.

In industry, magnesium is much the most important of the metals. The magnesium chloride for the electrolysis is obtained from brines of the type mentioned in Section 23, or even from seawater. The Dead Sea, where Israel built new electrolytic plants during the 1990s, is an attractive source (Figure 28.5). A typical electrolytic cell for magnesium production is shown in Figure 28.6.

Figure 28.5
A satellite photograph showing the Dead Sea which is fed from the north by the Jordan River (A). With no outflow, evaporation leads to a concentration of dissolved salts which is six times that of seawater. To the south, region B is divided into salt evaporators, from which magnesium chloride and other useful salts are obtained.

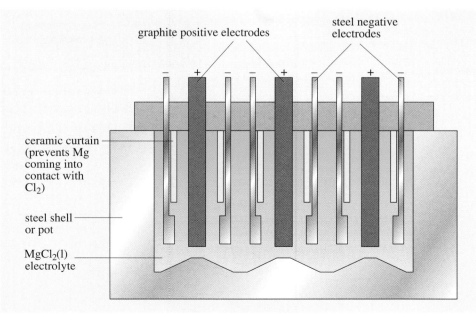

Figure 28.6
The production of magnesium. The melt temperature is about 700 °C. Molten magnesium is liberated at the negative electrodes, and rises to the surface where it is skimmed off and cast into ingots.

With a density of only $1.74\,\mathrm{g\,cm^{-3}}$, magnesium is the lightest structural metal, and this property is retained when it is alloyed with other metals, such as aluminium, to increase its strength and heat resistance. Magnesium alloys are therefore used extensively in aircraft construction. Other uses include lightweight car engine cases, and luggage. In 1999, world production of magnesium was 380 000 tonnes.

Some physical properties of the metals, their atoms and their ions are shown in Table 28.1.

Table 28.1 Some properties of Group II metals, atoms and ions

	Beryllium	Magnesium	Calcium	Strontium	Barium
density/g cm^{-3}	1.85	1.74	1.55	2.63	3.62
melting temperature/°C	1 287	649	839	768	727
$\Delta H^{\ominus}_{\mathrm{atm}}$(M, s)/kJ mol^{-1}	324	148	178	164	180
$E^{\ominus}(M^{2+}\,|\,M)$/V	−1.97	−2.36	−2.87	−2.90	−2.91
electronic configuration	[He]2s^2	[Ne]3s^2	[Ar]4s^2	[Kr]5s^2	[Xe]6s^2
I_1/kJ mol^{-1}	899	738	590	549	503
I_2/kJ mol^{-1}	1 757	1 451	1 145	1 064	965
I_3/kJ mol^{-1}	14 848	7 733	4 912	4 207	3 500
$r(M^{2+})$/pm	59	86	114	132	149

Note that in each case, the density, melting temperature and value of $\Delta H^{\ominus}_{\mathrm{atm}}$(M, s) are greater than the values for the adjacent alkali metal in the Periodic Table (compare Table 23.2).

The very negative values of $E^{\ominus}(M^{2+}\,|\,M)$ suggest that the alkaline earth metals are reducing agents of a strength comparable with that of the alkali metals. However, kinetic factors often make their reducing properties less spectacular. All of the metals dissolve readily in dilute acids with evolution of hydrogen:

$$M(s) + 2H^+(aq) = M^{2+}(aq) + H_2(g) \tag{28.5}$$

The reactions with water are more various. With calcium, strontium and barium, the reaction is brisk and hydrogen is evolved; for example

$$Ca(s) + 2H_2O(l) = Ca^{2+}(aq) + 2OH^-(aq) + H_2(g) \tag{28.6}$$

In the case of calcium, the calcium and hydroxide ions soon accumulate to a concentration at which the sparingly soluble calcium hydroxide, $Ca(OH)_2$, is precipitated.

The reaction of magnesium with water, however, is very slow and that of beryllium is negligible. This is because these two elements form relatively insoluble oxides and hydroxides which create a protective film on the metal surface.

In general, however, the reactions that you have met show that the Group II metals are readily oxidized to substances containing ions with a charge of +2. These ions have a noble gas configuration. However, oxidation does not proceed further to substances containing ions with a charge of +3 or more.

⬤ What term in Table 28.1 is relevant to this failure to observe further oxidation?

⬤ The third ionization energies, I_3, of the Group II elements are enormous. It is extremely difficult to remove a third electron from a Group II atom.

The very high values of I_3 are most clearly demonstrated by a comparison of the first three ionization energies of magnesium with those of the adjacent Group III metal, aluminium (Table 28.2).

Table 28.2 Ionization energies of magnesium and aluminium

	I_1/kJ mol^{-1}	I_2/kJ mol^{-1}	I_3/kJ mol^{-1}
magnesium	738	1 451	7 733
aluminium	578	1 817	2 745

⬤ Why is I_3 for a Group II atom such as magnesium so large?

⬤ The third ionization energy, I_3, is the ionization energy of the ion, Mg^{2+}. This ion has the noble gas configuration of the neon atom. On ionization, an electron must be removed from the inner shell (principal quantum number $n = 2$), whose electrons are much closer to the nuclear charge.

QUESTION 28.1

The Group II metals have higher melting temperatures, higher enthalpies of atomization and higher densities than their alkali metal neighbours (Figure 28.7) of the same Period. Explain this in terms of the electron gas model of metallic bonding.

QUESTION 28.2

Strontium and barium react with water, evolving hydrogen gas. In this they resemble rubidium and caesium. The reactions of the Group II elements, however, are much slower than those of the adjacent Group I elements. Suggest a reason for this.

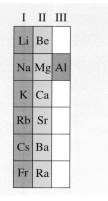

Figure 28.7
Relative positions of the alkali metals, the Group II elements and aluminium in the Periodic Table.

28.2 Why are there no Group II monohalides?

As noted in Section 28.1, the Group II elements cannot form ionic trihalides such as MgF_3, because the third ionization energies of these elements are so large. But a comparison of the first ionization energies of, say, potassium and calcium, or rubidium and strontium (Table 28.3) shows no enormous differences. Why then cannot the Group II elements form solid monohalides, $MX(s)$, like the alkali metals? To answer this question, we shall compare the cases of KCl and CaCl in particular detail. We begin with the Born–Haber cycle for a metal chloride, $MCl(s)$ (Figure 28.8). From this, we obtain the equation

$$\Delta H_f^{\ominus}(MCl, s) = \Delta H_{atm}^{\ominus}(M, s) + I_1(M) + \tfrac{1}{2}D(Cl-Cl) - E(Cl) + L(MCl, s) \qquad (28.7)$$

Table 28.3 First ionization energies of some Group I and Group II elements

M	$\dfrac{I_1(M)}{\text{kJ mol}^{-1}}$	M	$\dfrac{I_1(M)}{\text{kJ mol}^{-1}}$
K	419	Ca	590
Rb	403	Sr	549

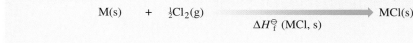
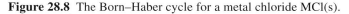

Figure 28.8 The Born–Haber cycle for a metal chloride MCl(s).

Table 28.4 shows some of the quantities that must be used in this equation when it is applied to KCl and CaCl.

Table 28.4 Terms in the Born–Haber cycles for potassium and calcium monohalides

Term	KCl	CaCl
$\Delta H^{\ominus}_{\text{atm}}(\text{M, s})/\text{kJ mol}^{-1}$	89	178
$I_1(\text{M})/\text{kJ mol}^{-1}$	419	590
$\frac{1}{2}D(\text{Cl}-\text{Cl})/\text{kJ mol}^{-1}$	122	122
$-E(\text{Cl})/\text{kJ mol}^{-1}$	−349	−349
$L(\text{MCl, s})/\text{kJ mol}^{-1}$	−718	−718
$\Delta H^{\ominus}_{\text{f}}(\text{MCl, s})/\text{kJ mol}^{-1}$	−437	−177

● Use Equation 28.7 and the data in Table 28.4 to calculate the lattice energy of KCl, $L(\text{KCl, s})$.

● $L(\text{KCl, s}) = -718\,\text{kJ mol}^{-1}$; the exercise is the one carried out on p. 124: the top five numbers in column 2 must add up to the bottom number.

As you can see, there is no value of $\Delta H^{\ominus}_{\text{f}}(\text{CaCl, s})$ in Table 28.4. This is because the compound has never been made, so its enthalpy of formation has not been determined experimentally. Let's try to obtain a value by other means. If we think in terms of ionic compounds, KCl(s) contains K^+ and Cl^- ions, and CaCl(s) would contain Ca^+ and Cl^- ions. Calcium is adjacent to potassium in the Periodic Table, and the ion Ca^+ has the same charge, and only one more proton and one more electron than K^+. Let us therefore assume that the two ions have similar sizes.

Now, the lattice energy of an ionic solid is given approximately by the Kapustinskii equation:

$$L = -\frac{WvZ_+Z_-}{r_+ + r_-} \qquad (20.11)$$

For both KCl and CaCl, $v = 2$, $Z_+ = 1$, $Z_- = 1$ and $r_- = r(Cl^-)$. We have also assumed that $r(Ca^+) = r(K^+)$.

⬤ What does this suggest about the lattice energies of CaCl(s) and KCl(s)?

⬤ They would be identical: $L(CaCl, s) = L(KCl, s) = -718\,kJ\,mol^{-1}$.

You can therefore put this value of $L(CaCl, s)$ into the last column of Table 28.4, and combine it with the other data to obtain a value of $\Delta H_f^\ominus(CaCl, s)$.

⬤ Do this now.

⬤ $\Delta H_f^\ominus(CaCl, s) = -177\,kJ\,mol^{-1}$, the sum of the top five numbers in the column.

The fact that this is a negative value is extremely interesting. It suggests that CaCl(s) might well be stable with respect to the elements of which it is composed. Thus, the reverse of the formation reaction, the decomposition into calcium and chlorine, is endothermic:

$$CaCl(s) = Ca(s) + \tfrac{1}{2}Cl_2(g); \ \Delta H_m^\ominus = 177\,kJ\,mol^{-1} \qquad (28.8)$$

To check whether CaCl(s) is thermodynamically stable to this reaction, we need the value of $\Delta S_m^\ominus$ as well as that of $\Delta H_m^\ominus$. As the entropy changes for analogous reactions are similar (Section 16), $\Delta S_m^\ominus$ can be estimated by using the value for the analogous reaction

$$KCl(s) = K(s) + \tfrac{1}{2}Cl_2(g) \qquad (28.9)$$

From the *Data Book*,

$$\Delta S_m^\ominus = S^\ominus(K, s) + \tfrac{1}{2}S^\ominus(Cl_2, g) - S^\ominus(KCl, s)$$
$$= 93.2\,J\,K^{-1}\,mol^{-1}$$

Using this as an estimate of $\Delta S_m^\ominus$ for Reaction 28.8,

$$\Delta G_m^\ominus = \Delta H_m^\ominus - T\Delta S_m^\ominus \qquad (28.10)$$
$$= 177\,kJ\,mol^{-1} - (298.15\,K \times 93.2\,J\,K^{-1}\,mol^{-1})$$
$$= 177\,kJ\,mol^{-1} - 28\,kJ\,mol^{-1}$$
$$= 149\,kJ\,mol^{-1}$$

This tells us that CaCl(s) is thermodynamically stable with respect to its constituent elements at 298.15 K, and that $\Delta G_f^\ominus(CaCl, s) = -149\,kJ\,mol^{-1}$. In other words, when we mix calcium metal and chlorine at room temperature, the formation of CaCl(s) is thermodynamically favourable. So why has no one succeeded in making it?

One possibility is that the reaction hasn't been properly tried out, and that *we* may be able to do better! But before we try the experiment, there is another possibility we should think about. It may be that although CaCl(s) is stable with respect to its elements, it is unstable to some other sort of decomposition.

⬤ Suggest such a decomposition.

● Since calcium so readily forms a dichloride, $CaCl_2$, one possibility is

$$2CaCl(s) = Ca(s) + CaCl_2(s) \qquad (28.11)$$

In this reaction, some of the Ca^+ ions in CaCl undergo reduction to calcium metal, and others are oxidized to Ca^{2+} ions in $CaCl_2$. This process of simultaneous oxidation and reduction is called **disproportionation**.

● Use our estimated value of $\Delta G_f^\ominus(CaCl, s)$ and the *Data Book* value $\Delta G_f^\ominus(CaCl_2, s)$ $= -748 \text{ kJ mol}^{-1}$ to estimate $\Delta G_m^\ominus$ for Reaction 28.11.

● $\Delta G_m^\ominus = \Delta G_f^\ominus(CaCl_2, s) - 2\Delta G_f^\ominus(CaCl, s)$

$\qquad = -748 \text{ kJ mol}^{-1} - 2 \times (-149 \text{ kJ mol}^{-1})$

$\qquad = -450 \text{ kJ mol}^{-1}$

Thus, although solid Group II monohalides are thermodynamically stable with respect to their elements at 25 °C, they are thermodynamically unstable with respect to disproportionation into the metal and MCl_2, which contains an M^{2+} ion with the noble gas configuration. This is the reason why such compounds do not exist.

QUESTION 28.3

Since the ion Ca^+ contains one more electron than K^+, you may be unhappy with the assumptions used in Section 28.2, namely that the ionic radii of K^+ and Ca^+, and therefore the lattice energies of KCl and CaCl, are of similar sizes. Suppose instead, therefore, that Ca^+ has the same radius as Rb^+, the next largest alkali metal cation. Given that the lattice energy of RbCl is -692 kJ mol^{-1}, reassess the stability of CaCl(s) using the method of Section 28.2. Does the change affect the conclusions in any serious way?

QUESTION 28.4

The alkali metals do not form dihalides. Table 28.5 contains the thermodynamic data that are needed to calculate the lattice energy of $BaF_2(s)$. By assuming that the ionic radii of Ba^{2+} and Cs^{2+} are identical, estimate $\Delta H_f^\ominus(CsF_2, s)$. Is your value consistent with the difficulties experienced in trying to make CsF_2? You should refer to the *Data Book* in answering this question.

Table 28.5 Terms in the Born–Haber cycles for barium and caesium difluorides

Term	BaF$_2$	CsF$_2$
$\Delta H_{atm}^\ominus(M, s)/\text{kJ mol}^{-1}$	180	76
$I_1(M)/\text{kJ mol}^{-1}$	503	376
$I_2(M)/\text{kJ mol}^{-1}$	965	2 236
$D(F-F)/\text{kJ mol}^{-1}$	158	158
$-2E(F)/\text{kJ mol}^{-1}$	-656	-656
$L(MF_2, s)/\text{kJ mol}^{-1}$	-2357	-2357
$\Delta H_f^\ominus(MF_2, s)/\text{kJ mol}^{-1}$	-1 207	-167

28.3 Group II in industry: lime and its applications

Lime is a general term, which includes quarried calcium carbonate, $CaCO_3$, and two other compounds that are easily obtained from it: **quicklime** or calcium oxide, CaO, and **slaked lime** or calcium hydroxide, $Ca(OH)_2$. When chalk or limestone is roasted at about $1\,200-1\,400\,°C$ (Figure 28.9), it loses carbon dioxide, leaving quicklime as a white powder:

$$CaCO_3(s) = CaO(s) + CO_2(g) \qquad (28.12)$$

Figure 28.9 The chalk-bearing countryside of England is scattered with the remains of old limekilns. This picture shows the brick pot in which alternate layers of chalk and burning coke produced quicklime.

If water is added to quicklime, a vigorous reaction occurs with evolution of heat: the product is slaked lime:

$$CaO(s) + H_2O(l) = Ca(OH)_2(s) \qquad (28.13)$$

It is also customary to speak of **dolomitic lime**, which describes dolomite, $CaMg(CO_3)_2$, and the products obtained by treating this substance in the same way as $CaCO_3$. Lime has the power to neutralize acidic substances. $Ca(OH)_2$, for example, is sparingly soluble in water ($1.3\,g$ litre^{-1} at $25\,°C$), and so both $Ca(OH)_2$ and — because of Equation 28.13 — CaO give alkaline solutions in water:

$$Ca(OH)_2(s) = Ca^{2+}(aq) + 2OH^-(aq) \qquad (28.14)$$

The hydroxide ions can then neutralize $H^+(aq)$:

$$H^+(aq) + OH^-(aq) = H_2O(l) \qquad (28.15)$$

The corresponding neutralization reaction of $CaCO_3$ was given at the beginning of Section 28:

$$CaCO_3(s) + H^+(aq) = Ca^{2+}(aq) + HCO_3^-(aq) \qquad (28.3)$$

This equation, however, is appropriate only at moderate acidities, such as those found in rainwater (around pH 5). If the acidity is high, the hydrogen carbonate ion reacts further:

$$HCO_3^-(aq) + H^+(aq) = H_2O(l) + CO_2(g) \qquad (28.16)$$

- Write the overall reaction for the neutralizing effect of $CaCO_3(s)$ in high acidities.

- It is the sum of Equations 28.3 and 28.16, which is:

$$CaCO_3(s) + 2H^+(aq) = Ca^{2+}(aq) + CO_2(g) + H_2O(l) \qquad (28.17)$$

Very large amounts of lime are consumed in neutralizing acidity. Pulverized limestone, slaked lime and quicklime are all used, where necessary, to lower the acidity of soils. Lime is also used to alleviate the effects of acid rain, which is caused by atmospheric reactions of the sulfur dioxide and oxides of nitrogen produced in coal or oil-fired power stations; for example

$$SO_2(g) + H_2O(l) + \frac{1}{2}O_2(g) = 2H^+(aq) + SO_4^{2-}(aq) \qquad (28.18)$$

You can learn more about this use of lime in the Case Study 'Acid Rain: Sulfur and Power Generation' in *Elements of the p Block*[3].

The chemical basis of other important uses of lime is revealed by regarding Reaction 28.12 as an equilibrium system:

$$CaCO_3(s) \rightleftharpoons CaO(s) + CO_2(g) \qquad (28.12)$$

The equilibrium lies well to the right at about $1\,200\,°C$; $CO_2(g)$ is driven off, and at this temperature $CaO(s)$ is the solid remaining. However, at ordinary temperatures, the equilibrium lies to the left, and both $CaO(s)$ and $Ca(OH)_2(s)$ will combine with $CO_2(g)$ to form $CaCO_3$; for example

$$CaO(s) + CO_2(g) = CaCO_3(s) \qquad (28.19)$$

In this reaction, the oxide ion from CaO combines with the non-metal oxide CO_2 to form a new anion, CO_3^{2-}:

$$Ca^{2+}O^{2-} + O{=}C{=}O = Ca^{2+}\begin{bmatrix} {}^-O \\ C{=}O \\ {}_-O \end{bmatrix} \qquad (28.20)$$

● Is there any change in the number of bonds formed by carbon in this reaction?

● No; the carbon–oxygen double bond in CO_2 is replaced by two $C{-}O^-$ single bonds.

Here then, a non-metallic atom exchanges two bonds formed with atomic oxygen for two bonds formed to oxygens bearing a negative charge. This is a general characteristic of some important reactions of lime. For example, SiO_2, or sand, is a network of Si—O single bonds (Figure 28.10). At high temperatures, calcium oxide will combine with it, opening up some of the Si—O—Si linkages, and replacing them with two Si—O$^-$ bonds (highlighted in colour in Figure 28.11).

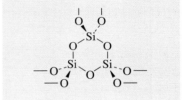

Figure 28.10
The essential features of the structure of silica, SiO_2. The compound consists of a network of Si—O single bonds. Each silicon atom forms four of these bonds, which are tetrahedrally disposed. Each oxygen atom forms two, and the ∠SiOSi bond angle is about 145°.

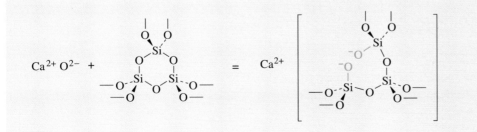

Figure 28.11 A representation of the structural changes that take place when calcium oxide reacts with silica at high temperatures. Each oxide ion attacks the silica structure, converting two Si—O bonds into two Si—O$^-$ bonds.

Again, two bonds formed by the non-metal with oxygen atoms are replaced by two bonds formed to oxygens that bear negative charges.

At temperatures above 1 000 °C, the product of the reaction in Figure 28.11 is a melt consisting of Ca^{2+} ions and anions composed of Si—O networks and Si—O⁻ bonds. This is exploited in the glass industry, which has a very long history (Figure 28.12).

Figure 28.12
The famous Portland vase is made from Roman glass of the early 1st century AD. It is violet–black, overlaid with white glass, into which figures are cut in cameo relief. The Duke of Portland lent it to the British Museum, where it was vandalized and completely shattered in 1845. A remarkably skilful reconstruction has left little trace of this disaster. In 1945, the Museum bought the vase outright.

Most manufactured glass is of the soda-lime variety. Sodium carbonate and limestone are heated above 1 000 °C with sand. At such high temperatures, the carbonates behave like the combinations $(Na_2O + CO_2)$ and $(CaO + CO_2)$. The carbon dioxide gas leaves the furnace, and the oxide anions open up Si—O⁻ sites on the silica network in reactions of the type shown in Figure 28.11. The result is a liquid in which Na^+ and Ca^{2+} ions are coordinated mainly by the O⁻ sites on Si—O network anions. On cooling, the liquid becomes progressively more viscous, forming the glass of everyday experience (Box 28.1).

A similar reaction to the one described in Box 28.1 occurs in the blast furnace, where lime removes silica and other non-metallic impurities into a liquid slag that floats above the molten iron. It should be clear from these applications alone that lime is a very large tonnage chemical; world consumption of limestone rock amounts to hundreds of millions of tonnes per annum.

Like the other alkaline earth metal carbonates, $CaCO_3$ has a very low solubility in water. It is, for example, precipitated when solutions containing $Ca^{2+}(aq)$ are treated with sodium carbonate solutions:

$$Ca^{2+}(aq) + CO_3^{2-}(aq) = CaCO_3(s) \qquad (28.21)$$

BOX 28.1 The molecular structure of glass

The structure shown in Figure 28.13a is that of a hypothetical non-metallic oxide, G_2O_3, in which each G atom forms three triangularly disposed G—O single bonds, each oxygen atom forming two linear bonds of this type. The structure is regular and, for the sake of simplicity, is assumed to be a flat sheet. When the oxide is heated with Na_2CO_3 and $CaCO_3$, carbon dioxide gas is lost, and Na_2O and CaO transfer their oxide ions to the G_2O_3 network, replacing G—O—G units with two G—O$^-$ bonds. This gives the irregular extended anion structure shown in Figure 28.13b, in which the Na^+ and Ca^{2+} ions are scattered and coordinated by oxygen atoms and O$^-$ sites. The same processes occur when glass is made from Na_2CO_3, $CaCO_3$ and SiO_2, but here the non-metallic oxide has the three-dimensional structure of Figure 28.10.

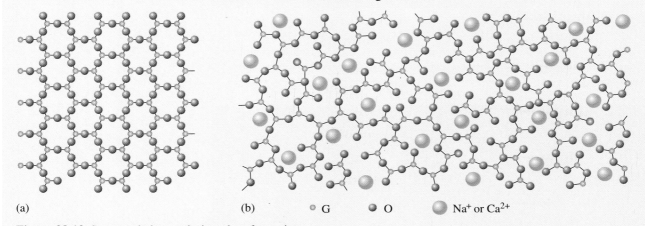

(a) (b) ○ G ● O ⬤ Na^+ or Ca^{2+}

Figure 28.13 Structural change during glass formation.

The white suspensions of insoluble carbonate, however, will dissolve if carbon dioxide is bubbled through the solution. This is a special case of Equation 28.3, in which CO_2 acts as the source of acidity:

$$CaCO_3(s) + H_2O(l) + CO_2(g) \rightleftharpoons Ca^{2+} + 2HCO_3^-(aq) \qquad (28.22)$$

If, however, one tries to make solid $Ca(HCO_3)_2$ from the resulting clear solution by evaporation, the reactants are reformed because $Ca(HCO_3)_2$ is very unstable to decomposition:

$$Ca(HCO_3)_2(s) = CaCO_3(s) + CO_2(g) + H_2O(l) \qquad (28.23)$$

Equation 28.22 is the key to some important natural phenomena. Because of dissolved atmospheric CO_2, natural water percolating through limestone can become saturated with dissolved calcium hydrogen carbonate. The dissolved calcium is an important contribution to the hardness of such water. However, boiling eliminates this hardness by returning equilibrium in Reaction 28.22 to the left and destroying the dissolved hydrogen carbonate. A similar reversion to insoluble calcium carbonate may result in the formation of striking stalagmites and stalactites from water droplets in limestone caverns (Figures 28.14 and 28.15).

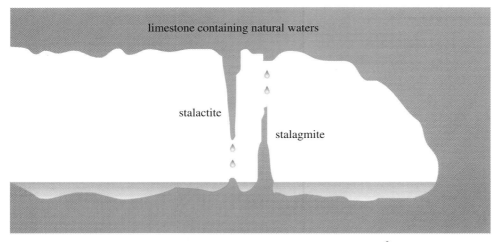

Figure 28.14 The evaporation of dripping water containing dissolved $Ca^{2+}(aq)$ and $HCO_3^-(aq)$ on surface irregularities in limestone caverns can shift the equilibrium in Reaction 28.22 to the left, and give rise to needle-like structures called stalagmites and stalactites. Stalactites are needles of limestone suspended from the roof; stalagmites rise up from the floor.

Figure 28.15 Stalagmites and stalactites in a limestone cave.

QUESTION 28.5

The highest normal oxide of phosphorus has the empirical formula P_2O_5, and contains discrete P_4O_{10} molecules (Structure **28.1**). Calcium oxide reacts with this oxide to give calcium phosphate, $Ca_3(PO_4)_2$:

$$6CaO(s) + P_4O_{10}(s) = 2Ca_3(PO_4)_2(s) \qquad (28.24)$$

in which the phosphate ion, PO_4^{3-}, has Structure **28.2**. Show how the kind of structural change that occurs when CaO reacts with CO_2 and SiO_2, as in Reaction 28.20 and the reaction shown in Figure 28.11, also takes place here.

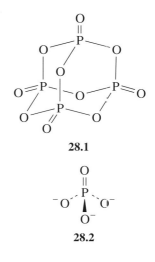

28.1

28.2

28.4 Oxides, hydroxides and carbonates of the Group II elements

If the Group II metals are heated in air at 600 °C, beryllium, magnesium, calcium and strontium form normal oxides, MO. Barium forms a peroxide, BaO_2; however, if the temperature is increased to 800 °C, the peroxide decomposes to BaO. Beryllium oxide crystallizes with the wurtzite structure (Figure 28.16); the coordination number of beryllium is four. All of the other oxides have the NaCl structure, in which the alkaline earth metal ion is six coordinate.

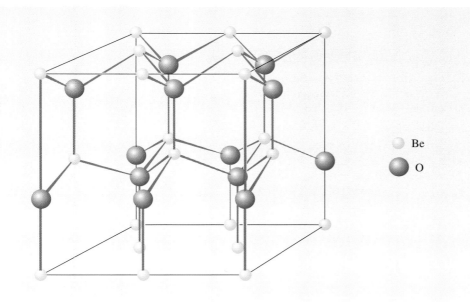

Be
O

Figure 28.16 The wurtzite structure of BeO. Each beryllium is tetrahedrally surrounded by four oxygens, and each oxygen is tetrahedrally surrounded by four berylliums.

The hydroxides $Be(OH)_2$, $Mg(OH)_2$ and $Ca(OH)_2$ have very low solubilities in water, and can be precipitated by adding sodium hydroxide to solutions containing the aqueous ions. They will dissolve in hydrochloric acid and nitric acid, but $Be(OH)_2$ alone will also dissolve in strongly *alkaline* solutions, forming the complex $[Be(OH)_4]^{2-}(aq)$ (Structure **28.3**):

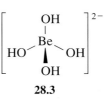

28.3

$$Be(OH)_2(s) + 2OH^-(aq) = [Be(OH)_4]^{2-}(aq) \tag{28.25}$$

You will recall that oxides or hydroxides that neutralize both acids and bases are called *amphoteric*.

Both the hydroxides and the carbonates decompose to the oxides when heated:

$$M(OH)_2(s) = MO(s) + H_2O(g) \tag{28.26}$$

$$MCO_3(s) = MO(s) + CO_2(g) \tag{28.27}$$

Table 28.6 shows the decomposition temperatures and $\Delta H_m^{\ominus}$ values for the carbonate reaction. They increase down the Group.

Table 28.6 Values of $\Delta H_m^\ominus$ (at 298.15 K) and the decomposition temperature for the decomposition reaction of Group II carbonates

	Mg	Ca	Sr	Ba
$\Delta H_m^\ominus$ /kJ mol^{-1}	101	178	235	269
decomposition temperature/K	580	1 110	1 370	1 570

QUESTION 28.6

Explain why only barium forms a peroxide when the metals Be, Mg, Ca, Sr, and Ba are strongly heated in oxygen.

QUESTION 28.7

By using a thermodynamic cycle of the type shown in Figure 25.6, explain why the stability of the alkaline earth metal carbonates increases down the Group.

QUESTION 28.8

Magnesium undergoes little or no reaction with water or with sodium hydroxide solution; beryllium does not react with water, but it dissolves readily in sodium hydroxide solution with evolution of hydrogen. Explain this, and write an equation that includes aqueous hydroxide ions for the reaction of beryllium metal with water.

28.5 Are the Group II dihalides ionic?

The Group II dihalides can be made by the reactions of the metals with the halogens. Up till now, we have treated the bonding in Group II compounds and solution as ionic. The dihalides, however, do not altogether fit into this framework.

With the exception of MgF_2, CaF_2, SrF_2 and BaF_2, which are sparingly soluble, all the dihalides dissolve readily in water to give conducting solutions:

$$MX_2(s) = M^{2+}(aq) + 2X^-(aq) \tag{28.28}$$

So far, so good. However, in the structure of an ionic solid, each supposed ion should be surrounded by ions of opposite charge. Table 28.7 shows that four of the dihalides have CdI_2 or $CdCl_2$ layer structures, in which the supposed halide ions are coordinated on one side by three metal ions in the same layer, and on the other side by other halide ions in an adjacent layer (Figure 28.17). This is not what we expect of a solid that is composed of ions.

Table 28.7 Structure of some alkaline earth metal halides

	F	Cl	Br	I
Mg		CdCl$_2$ or CdI$_2$ layer structures		
Ca	three-dimensional structures, such as rutile, fluorite and other types, often described as assemblies of ions			
Sr				
Ba				

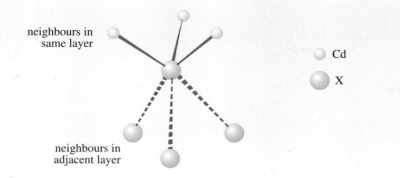

neighbours in
same layer

Cd

X

neighbours in
adjacent layer

Figure 28.17
The environment of the halogen in a $CdCl_2$ or CdI_2 layer structure. Each layer consists of three decks, the central deck containing cadmiums, and the upper and lower decks containing halogens. On one side of each halogen, there are three cadmiums in the same layer at a relatively short distance. On the other side, there are three halogens in the next layer at a considerably larger distance.

In Table 28.7 these troublesome layer structures are printed in green at the top right; you know of two distinct ways of explaining this. One was to argue that the electronegativity difference between the Group II element and the halogen is least at the top right of the Table, so it is here that the ionic model is most likely to break down.

⬤ What is the second type of explanation?

⬤ At the top right of Table 28.7, small cations such as Mg^{2+} are combined with large anions such as I^-. The small cation may then polarize the electron cloud of the large anion, introducing a degree of covalency.

Let's use these explanations to make predictions about beryllium halides, which have not been included in Table 28.7.

⬤ Will beryllium dihalides conform less closely or more closely to an ionic model than magnesium dihalides?

⬤ Less closely; using the first argument, beryllium is more electronegative than magnesium, so there will be smaller electronegativity differences in its halides than for magnesium halides. Using the second argument, Be^{2+} is smaller than Mg^{2+}, and will cause more polarization of halide ions.

This prediction gives a correct result. Unlike other Group II dichlorides and di-fluorides, molten $BeCl_2$ and BeF_2 are very poor conductors of electricity. Moreover, the structure of solid $BeCl_2$ is even less consistent with an ionic picture than the layer structures of $MgCl_2$, $MgBr_2$ and MgI_2. It consists of parallel chains in which each beryllium is tetrahedrally coordinated to four chlorines (Figure 28.18). Each chlorine has two beryllium atoms coordinated to it, but in all other directions it is surrounded by chlorine atoms in other chains, which are considerably further away.

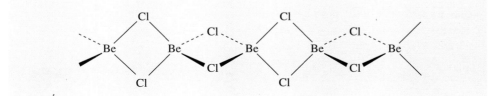

Figure 28.18 The chain structure of solid $BeCl_2$.

BeF$_2$ has a quartz-like structure in which beryllium is again tetrahedrally coordinated. However, the twofold coordination of fluorine is not linear as one might expect in a collection of ions: the $\angle$BeFBe bond angle is about 145°.

QUESTION 28.9

Sodium chloride has a structure in which each sodium is surrounded by six chlorines, and each chlorine is surrounded by six sodiums. Magnesium is sodium's neighbour in the Periodic Table, and its chloride has a CdCl$_2$ layer structure. Use the arguments of Section 28.5 to explain this difference.

28.6 Complexes of Group II elements

The Group II elements are less willing to form new complexes in solution than the transition elements, but more willing than the alkali metals. Like the alkali metals, for example, their cations complex with polycyclic ethers and cryptands (Structures **27.1**, **27.2**, **27.5** and **27.6**). Unlike the alkali metals, however, they form complexes in aqueous solution with the hexadentate ligand, edta^{4-} (Table 26.1):

$$M^{2+}(aq) + edta^{4-}(aq) = [M(edta)]^{2-}(aq) \tag{28.29}$$

The structure of the calcium–edta complex is shown in Figure 28.19; notice how the edta^{4-} ligand wraps itself around the metal ion so that the latter is octahedrally coordinated by donating atoms with non-bonded electron pairs.

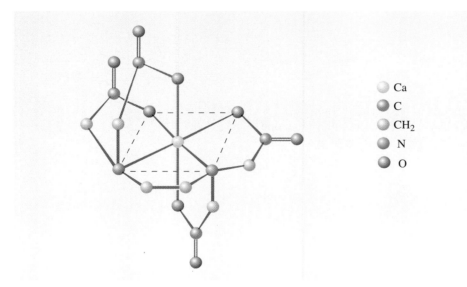

- Ca
- C
- CH$_2$
- N
- O

Figure 28.19 The structure of the complex [Ca(edta)]$^{2-}$.

The Group II element that forms complexes most readily with simple unidentate ligands like H$_2$O, NH$_3$, OH$^-$ and halide ions, is beryllium. One mark of this that you have already encountered is the formation of the complex [Be(OH)$_4$]$^{2-}$, which shows the amphoteric character of Be(OH)$_2$ (Section 28.4). In this and other cases, beryllium shows a marked preference for tetrahedral coordination. Thus, if beryllium metal is dissolved in dilute hydrochloric or sulfuric acid, and if the solutions are evaporated until crystallization of solids begins, the products are the hydrates

$[Be(H_2O)_4]Cl_2$ and $[Be(H_2O)_4]SO_4$, respectively. In these compounds, the water of the tetrahedral complex $[Be(H_2O)_4]^{2+}$ is held very tenaciously. Likewise, if $BeCl_2$ is recrystallized from liquid ammonia, the product is $[Be(NH_3)_4]Cl_2$.

Other signs of this preference of beryllium for tetrahedral, fourfold coordination were the structures of BeO, $BeCl_2$ and BeF_2 (Sections 28.4 and 28.5). It can be linked to the failure of beryllium compounds to match the expectations of an ionic model. Consider, for example, the complex $[Be(NH_3)_4]^{2+}$. In Section 26.2, we said that if one adopts a covalent picture of complex formation, a metal–ligand bond involves the donation of non-bonded electron pairs on the nitrogen atom of the ammonia ligand to the central metal ion.

⬤ If so, what type of electronic configuration does the beryllium in $[Be(NH_3)_4]^{2+}$ have?

⬤ The configuration of neon, a noble gas: Be^{2+} has the configuration $1s^2$; the four NH_3 ligands donate four non-bonded pairs and these eight electrons will occupy the 2s and 2p levels of beryllium to give the configuration $1s^2 2s^2 2p^6$.

This explanation implies that beryllium's preference for four-coordination is a mark of covalent bond formation. It therefore links that preference with the structural evidence for the breakdown of an ionic model which we noted in Section 28.4.

28.7 Summary of Section 28

1   Magnesium and calcium are very abundant. They occur mainly as carbonate compounds such as limestone, or as chloride brines from which they can be obtained by evaporation. The metals are made by electrolysis of their molten chlorides.

2   The metals have higher densities, melting temperatures and enthalpies of atomization than the adjacent alkali metals in the Periodic Table. This is because they have two bonding electrons per metal atom rather than one. They are powerful reducing agents that react readily with acids to give hydrogen gas. The reaction with water is negligible or very slow in the cases of beryllium and magnesium because of insoluble, protective oxide/hydroxide films. However, calcium, strontium and barium react readily with water, yielding alkaline solutions and hydrogen.

3   In the atoms of Group II elements, the nuclear charge is lower than for other elements in the same Period (except for the alkali metals). The first two ionizations, which lead to a noble gas configuration, therefore have relatively low energies. By contrast, the third ionization energies are unusually large because an electron must be removed from an inner shell, whose electrons are closer to the nucleus than the outer s^2 electron pair.

4   The combination of relatively low first and second, and high third ionization energies is primarily responsible for the readiness of the Group II metals to form substances containing M^{2+} ions, but not ions of higher charge.

5   Group II substances containing M^+ ions are unknown because they are unstable to disproportionation into the metals and substances containing M^{2+} ions.

6   Limestone, $CaCO_3$, yields quicklime, CaO, on heating, and water then converts quicklime into slaked lime, $Ca(OH)_2$. All three compounds can neutralize acids or acidic oxides.

7   Limestone is used in glass manufacture. At high temperatures, carbon dioxide is lost, leaving calcium oxide, which transfers its oxide ion to non-metallic oxides such as SiO_2, replacing bonds formed with oxygen atoms by bonds formed with oxygens bearing a negative charge. With silica, this generates a silicate anion framework containing cations such as Ca^{2+} and Na^+. This melt becomes a glass on cooling.

8   Limestone is slightly soluble in rainwater (which is acidic because of dissolved CO_2). The consequent erosion creates surface features like cliffs and gorges, and subterranean pores, cracks and fissures in limestone strata, including stalactites and stalagmites. Consequently, these strata both store and transport water.

9   The ionic radii of the M^{2+} ions increase from beryllium to barium. Consequently, the thermal stabilities of the peroxides, hydroxides and carbonates, which decompose to oxides of larger $-L$ value, increase down the Group. Barium, for example, forms a peroxide when the metal is heated in air, but the Group II elements above it do not.

10   The Group II halides BeX_2, $MgCl_2$, $MgBr_2$, MgI_2 and CaI_2 have properties not wholly consistent with an ionic model. For example, $BeCl_2$ has a chain structure, and $MgCl_2$, $MgBr_2$ and MgI_2 have layer structures; molten BeF_2 and $BeCl_2$ are poor conductors of electricity. These properties can be attributed to polarization brought about by the juxtaposition of small cations and large anions, or to smaller electronegativity differences.

11   The Group II ions form complexes with polycyclic ethers and cryptands and with $edta^{4-}$ in aqueous solution. Beryllium is the Group II element that complexes most strongly with common unidentate ligands; in doing so, it takes on tetrahedral coordination. In terms of a covalent bonding model, this gives beryllium the electronic configuration of the subsequent noble gas, neon.

QUESTION 28.10

Boilers often become lined with 'scale', which is deposited from hot water containing dissolved salts. Scale typically consists of sparingly soluble calcium compounds, such as $CaCO_3$ and $CaSO_4$. One way of removing it is to treat the inside of the boiler with a solution of sodium ethylenediaminetetraacetate. Explain.

QUESTION 28.11

Beryllium chloride, $BeCl_2$, dissolves in ethylenediamine (Structure **26.2**), forming a beryllium complex. If solvent is evaporated, the chloride of this complex eventually crystallizes out. Predict the formula and structure of the complex chloride.

APPENDIX: THERMODYNAMICS IN THIS BOOK

The thermodynamics in this Book has been put together for a particular purpose — an exploration of the reactions of metals. For this reason, it is less general than other accounts of the subject, and in this Appendix, comments are made on some of the important differences.

1   The chemical reactions that are discussed in this Book take place mainly at constant temperature and constant pressure. Under these conditions, the enthalpy change, ΔH, and the Gibbs energy change, ΔG, are the energy changes that are especially informative. They therefore receive unusual emphasis, and the internal energy function, U, is not discussed.

2   At constant pressure, ΔH is informative because for both a reaction and a phase change of a pure substance, it can be equated to the heat absorbed from the surroundings:

$$\Delta H = q \text{ (at constant } p) \tag{6.1}$$

In the constant-pressure processes to which Equation 6.1 applies, some work may be done alongside the absorption or evolution of heat. This work arises from volume changes. For example, when water boils at one atmosphere pressure, the volume expands some 1 700 fold. The boiling water must create the space for this expansion by pushing back the surrounding atmosphere that presses in on it. This means that the boiling water must do work, akin to the work that must be done in blowing up a balloon. ΔH is an energy change that is defined so that Equation 6.1 remains correct when this type of work is done. In other words, Equation 6.1 holds provided that the only work done relates to volume change.

3   If other sorts of work are performed as well, then Equation 6.1 must include them. Thus, if electrical work, w_{el}, is performed on the substance under consideration, then Equation 6.1 becomes:

$$\Delta H = q + w_{el} \tag{8.1}$$

This is our chosen but restricted statement of the first law of thermodynamics. It applies to constant-pressure processes in which the only types of work that are done are electrical work, which appears explicitly in Equation 8.1, and the work of volume change discussed in point 2. We chose this statement of the first law because it is particularly suited to the discussion of the calorimetric measurements made in Sections 8–10.

Equation 8.1 implies that the enthalpy change of a reaction cannot be identified with the heat absorbed if electrical work is performed at the same time. Important examples of this are the self-driving electrical batteries discussed in the *Batteries* Case Study. These do electrical work for the benefit of the user, and the heat absorbed is then not equal to the enthalpy change of the reaction that takes place inside the battery.

4   ΔG is informative because, at constant temperature and pressure, its sign and
     size are an indication of how thermodynamically favourable a reaction is.
     Our decision to emphasize ΔG in this way gives the equilibrium constant, K,
     a very low profile in this Book. K can be related to $\Delta G_m^{\ominus}$ by Equation 11.7 or
     11.8; for example

$$\Delta G_m^{\ominus} = -2.303RT \log K \tag{11.8}$$

Note, however, that as Table 11.1 implies, the values of K obtained from $\Delta G_m^{\ominus}$
values using this equation are dimensionless. You may be used to equilibrium
constants with dimensions of concentration (e.g. mol litre^{-1} in Question 4.2); we do
not discuss the difference between these two kinds of equilibrium constant here. All
that we will say here is that, if the dimensions in the kind of equilibrium constant
that you are familiar with are removed, the resulting number is not very different
from the number obtained from $\Delta G_m^{\ominus}$ by applying Equation 11.8.

LEARNING OUTCOMES

Now that you have completed *Metals and Chemical Change*, you should be able to do the following things:

1 Recognize valid definitions of and use in a correct context the terms, concepts and principles in the following Table. (All Questions)

List of scientific terms, concepts and principles introduced in *Metals and Chemical Change*

Term	Page number	Term	Page number	Term	Page number
activity series	37	entropy	45	reducing agent	25
amphoteric oxide (hydroxide)	32	entropy change	47	reduction	19
analogous reactions	112	first ionization energy	123	second law of thermodynamics	45
adiabatically enclosed system	55	first law of thermodynamics	53	slaked lime	192
ball lightning	106	gas constant	85	standard e.m.f., $E^{\ominus}$	140
bidentate ligand	174	Gibbs function	84	standard enthalpy of atomization, $\Delta H^{\ominus}_{atm}$	123
bond dissociation energy	123	heat	52		
Born–Haber cycle	123	Hess's law	64	standard enthalpy of formation, $\Delta H^{\ominus}_{f}$	65
calorimeter	56	hydrate	170		
chlor-alkali industry	154	ionic model	129	standard molar enthalpy change, $\Delta H^{\ominus}_{m}$	65
complex	172	isolated system	45		
coordinating atom	173	Kapustinskii equation	136	standard molar Gibbs function change, $\Delta G^{\ominus}_{m}$	84
crown ether	176	kinetically stable	98		
cryptand	179	lattice energy, L	123	standard redox potentials	141
discharge mode	140	ligand	173	state function	55
disproportionation	191	lime	192	state of a system	55
dolomitic lime	192	Madelung constant, M	133	system	45
ductile	13	malleable	13	thermodynamic decomposition temperature	116
electrical work	53	oxidation	19	thermodynamically stable	98
electrochemical cell	140	oxidizing agent	25	third law of thermodynamics	79
electromotive force (e.m.f.), E	140	phase transition	56	unidentate ligand	174
electron affinity	123	polydentate ligand	174	work	53
Ellingham diagram	117	quicklime	192		
enthalpy of reaction	42	redox reactions	20		

①✗ 2   Identify instances of oxidation and reduction in terms of loss and gain of electrons. (Questions 2.1 and 2.4)

①✗ 3   Describe and, if appropriate, write equations for the effect of dilute acid on alkali metals, the alkaline earth metals, zinc, iron, copper, mercury, silver and gold. (Question 2.2)

①✗ 4   Use the classification of metals made in Sections 2–4 to predict the reaction (or lack of a reaction) when a metal in one class is treated with a halide, oxide or aqueous ion of a metal in a different class. (Questions 2.3 and 4.1)

②+ 5   Use criteria derived from the second law of thermodynamics to predict whether a given change should occur or not. (Questions 7.1–7.3)

6   For a change at constant pressure, express the first law of thermodynamics in terms of the enthalpy of the system, and hence determine the enthalpy change for a pure substance. (Question 8.1)

7   With reference to a constant-pressure calorimeter: (a) relate the experimental measurements made to the accompanying enthalpy changes; (b) relate the measurements in (a) to the enthalpy change at constant temperature for the process under consideration; (c) state any assumption underlying the procedures used. (Questions 8.2 and 8.5)

8   Given the results of a calorimetric experiment, calculate the corresponding molar enthalpy change. (Questions 8.2, 8.3 and 8.5)

②+③ 9   Show how the molar enthalpy change for a reaction depends on what multiple of the balanced reaction equation is used. (Questions 8.4–8.6 and 12.1)

10   Calculate entropies of fusion or vaporization from enthalpies of fusion or vaporization, and melting or boiling temperatures. (Question 10.1)

11   Use appropriate experimental data to determine entropy changes over a given temperature range. (Questions 10.2 and 10.3)

②+12   Define the Gibbs function in terms of enthalpy, entropy and temperature, and use this definition to derive an expression relating the standard molar change in the Gibbs function for a reaction, $\Delta G_m^{\ominus}$, to the corresponding standard molar enthalpy and entropy changes, $\Delta H_m^{\ominus}$ and $\Delta S_m^{\ominus}$, respectively. (Questions 11.1, 11.3 and 12.1)

②+③ 13   Write down the formation reaction for a given compound or aqueous ion. (Questions 9.1, 9.2, 11.5 and 12.2)

②+ 14   Given a table containing values of $\Delta H_f^{\ominus}$, $\Delta G_f^{\ominus}$ and $S^{\ominus}$ for elements, compounds and aqueous ions, calculate: (a) standard molar enthalpies of reaction; (b) standard molar entropies of reaction; (c) standard molar Gibbs function changes of reaction. (Questions 9.1–9.4, 10.4, 11.1, 11.3–11.5, 12.1–12.3, 13.2, 13.3, 15.1, 15.2 and 17.1)

①+② ✗ 15   Express the criterion for spontaneous change in terms of standard molar quantities, and hence use the results of the calculations in Learning Outcome 14 to predict whether or not a given reaction is thermodynamically favourable. (Questions 11.1, 11.3, 13.1–13.3, 15.3 and 17.2)

16   Given the appropriate thermodynamic data, analyse statements made about reactivity or lack of reactivity in terms of kinetic and thermodynamic effects. (Questions 13.2, 13.3 and 15.3)

17   Decide when similar reactions have similar entropy changes. (Questions 16.1 and 16.2)

18 Given a list of reactions involving gases and solids only, identify those that have: (a) large negative entropy changes; (b) large positive entropy changes; (c) small entropy changes, which may be either positive or negative. (Question 16.2).

19 Use values of $\Delta H_m^{\ominus}$ and $\Delta S_m^{\ominus}$ for a reaction at 298.15 K to estimate a value of $\Delta G_m^{\ominus}$ at higher temperatures. (Question 17.1)

20 Use Ellingham diagrams to predict temperatures at which reactions become thermodynamically favourable. (Question 17.2)

21 Relate the categorization of techniques of metal extraction given in Sections 3 and 17 to the position of metals in the Periodic Table, and to the thermodynamics of metals and their compounds. (Questions 17.3 and 18.4)

22 Construct and use Born–Haber cycles to calculate the lattice energies of metallic halides. (Questions 18.1–18.3 and 20.2)

23 Given the formula unit for an ionic compound, use appropriate theoretical expressions to calculate the lattice energy from the appropriate combination of the Madelung constant, ionic charges and ionic radii. (Questions 20.1–20.5)

24 Given a table of standard redox potentials, identify strong and weak oxidizing or reducing agents, and use the table to predict whether redox reactions are thermodynamically favourable or not. (Questions 21.1–21.3)

25 Explain and exploit the relationship that exists among the typical elements between electronic configurations, nuclear charges, position in the Periodic Table and ionization energies. (Question 22.1)

26 Recall important items of information about the Group I and Group II elements, notably those summarized in Sections 23.2, 24.3, 25.4, 26.3, 27.2 and 28.7, and combine them to gain new insights about the chemistry of those elements. (Questions 23.1, 25.1, 27.1, 28.1, 28.2, 28.5 and 28.8–28.11)

27 Use thermodynamic cycles to relate the relative stabilities and decomposition temperatures of Group I and Group II compounds to variations in cation size, and to the lattice energies of solid compounds involved in the decomposition reactions. (Questions 25.2, 25.3, 28.6 and 28.7)

28 Given the crystal structure of a solid containing a complex, identify the complex along with the ligands that it contains, and write equations for reactions in which the coordination of the central element in the complex is maintained, but the ligands are replaced by others. (Questions 26.1, 28.10 and 28.11)

29 Estimate the standard enthalpies of formation of unknown ionic compounds by inserting estimated values of their lattice energies into a Born–Haber cycle, and then search for thermodynamically favourable decomposition reactions that might explain why such compounds do not exist. (Questions 28.3 and 28.4)

QUESTIONS: ANSWERS AND COMMENTS

QUESTION 2.1 (Learning Outcome 2)

Reactions (i), (ii), (iii) and (v) are redox reactions.

(i)   Potassium is oxidized and hydrogen is reduced. The potassium atoms in the metal lose an electron and become $K^+(aq)$; the hydrogen ions gain an electron to become hydrogen atoms, which then combine to give $H_2(g)$.

(ii)   Copper ions are reduced and metallic iron is oxidized.

(iii) MgF_2 is an ionic compound formulated as $Mg^{2+}(F^-)_2$. Thus, magnesium atoms are oxidized to Mg^{2+}, and fluorine atoms in F_2 are reduced to F^-.

(v)   Chlorine atoms in Cl_2 molecules are reduced, and iron in the form $Fe^{2+}(aq)$ loses an electron and is oxidized to $Fe^{3+}(aq)$.

Note that with the ionic formulation $Ca^{2+}(F^-)_2$ for CaF_2, calcium and fluorine retain the ionic forms Ca^{2+} and F^- in reaction (iv). This is therefore not a redox reaction.

QUESTION 2.2 (Learning Outcome 3)

Rubidium is an alkali metal and strontium is an alkaline earth metal, so both react vigorously with aqueous hydrogen ions and therefore with hydrochloric acid. Hydrogen gas is produced, and Rb^+ or Sr^{2+} ions are formed in solution:

$$Rb(s) + H^+(aq) = Rb^+(aq) + \tfrac{1}{2}H_2(g) \qquad\qquad (Q.1)$$
$$Sr(s) + 2H^+(aq) = Sr^{2+}(aq) + H_2(g) \qquad\qquad (Q.2)$$

Silver was placed in our first class of metals, and so does not react.

QUESTION 2.3 (Learning Outcome 4)

Copper and silver are members of our first class of metals, iron and tin are members of the second, and magnesium and calcium are members of the third. A reaction therefore occurs in cases (ii) and (iii):

$$Sn(s) + 2Ag^+(aq) = Sn^{2+}(aq) + 2Ag(s) \qquad\qquad (Q.3)$$
$$Mg(s) + Sn^{2+}(aq) = Mg^{2+}(aq) + Sn(s) \qquad\qquad (Q.4)$$

but not in cases (i) and (iv).

QUESTION 2.4 (Learning Outcome 2)

(i)   Tin is oxidized; copper (in the form $Cu^{2+}(aq)$) is reduced.

(ii)   Lead is oxidized; silver is reduced.

(iii)   Bromine is oxidized; chlorine is reduced.

(iv)   Barium is oxidized; chlorine is reduced.

(v)   With the formulation Ag^+Cl^- for $AgCl$, it is apparent that there is no oxidation or reduction.

(vi)   Zinc is oxidized; silver is reduced.

(vii)   Potassium is oxidized; aluminium is reduced.

(viii)   Aluminium is oxidized; iron is reduced.

(ix)   Iron is oxidized and oxygen is reduced; using the ionic formulation, you can see that FeO contains Fe^{2+} and O^{2-} ions, whereas Fe_2O_3 contains Fe^{3+} and O^{2-} ions.

(x)   Iron is oxidized; chlorine is reduced. Notice that in both its compounds and in aqueous solution, iron forms more than one kind of monatomic ion. This is not true of many of the metals that we consider in this Book.

QUESTION 4.1 (Learning Outcome 4)

If the qualitative arguments built up in Sections 2–4 are correct, a reaction would be expected in all cases except (i), (iv) and (vi).

(ii) $Zn + CuO = ZnO + Cu$

(iii) $Zn + FeCl_2 = ZnCl_2 + Fe$

(v) $3K + AlCl_3 = 3KCl + Al$

(vii) $Zn(s) + 2Ag^+(aq) = Zn^{2+}(aq) + 2Ag(s)$

(viii) $Fe + PbBr_2 = FeBr_2 + Pb$

(ix) $Zn(s) + H_2O(g) = ZnO(s) + H_2(g)$

(x) $Ba + FeBr_2 = BaBr_2 + Fe$

All parts of this question except (iii), (v) and (viii) can be answered either by reference to the classes in Sections 2.2 and 2.3 or to the series in Section 4. For (iii), (v) and (viii), the two metals are in the same class, and only the series in Section 4 can be used.

QUESTION 4.2 (Learning Outcome 1)

The equilibrium constant, K, is given by

$$K = \frac{[H^+(aq)][Ac^-(aq)]}{[HAc(aq)]} \qquad (Q.5)$$

Because acetic acid is a weak acid, equilibrium lies to the left, and a solution of acetic acid contains only small concentrations of hydrogen and acetate ions; HAc(aq) is the predominant species in solution. This is marked by the fact that K is small; the actual value is 1.8×10^{-5} mol litre^{-1}.

QUESTION 6.1 (Learning Outcome 6)

The value of ΔH quoted in the question is negative, so the reaction is exothermic. According to the principle of the conservation of energy, the reverse reaction, the decomposition of one mole of NaCl, must absorb an equivalent amount of energy. Hence it must be *endothermic*; that is

$$NaCl(s) = Na(s) + \tfrac{1}{2}Cl_2(g); \Delta H = +411.2\,kJ \qquad (Q.6)$$

According to these values, solid sodium chloride is of lower enthalpy than a mixture of sodium metal and chlorine gas (at constant temperature and pressure), so Thomsen's hypothesis *could* explain the behaviour of this system.

QUESTION 7.1 (Learning Outcome 5)

This time, block 2 absorbs heat q, so $\Delta S_2 = q/T_2$, and block 1 loses heat q, so $\Delta S_1 = -q/T_1$. Thus

$$\Delta S_{total} = \frac{q}{T_2} - \frac{q}{T_1} \qquad (Q.7)$$

Taking the two blocks to be an isolated system, ΔS_{total} is negative because $T_2 > T_1$. This violates the second law: the process cannot occur.

QUESTION 7.2 (Learning Outcome 5)

For Reaction 5.2

$$Na(s) + \tfrac{1}{2}Cl_2(g) = NaCl(s) \qquad (5.2)$$

$\Delta S = -90.7 \text{ J K}^{-1}$ and $\Delta H = -411.2 \text{ kJ}$, so

$$\frac{\Delta H}{T} = -\frac{411.2 \text{ kJ}}{298.15 \text{ K}} = -1.379 \text{ kJ K}^{-1} = -1379 \text{ J K}^{-1}$$

Thus

$$\Delta S - \frac{\Delta H}{T} = \{-90.7 - (-1379)\} \text{ J K}^{-1}$$
$$= +1\,288 \text{ J K}^{-1}$$

The value is positive, so the second law is vindicated!

QUESTION 7.3 (Learning Outcome 5)

(i) In this case the reaction is endothermic, so heat is withdrawn from the surroundings; ΔS_{surr} must be negative:

$$\Delta S_{surr} = -\frac{\Delta H}{T} = -\frac{47100 \text{ J}}{298.15 \text{ K}}$$
$$= -158.0 \text{ J K}^{-1}$$

(ii) $\Delta S_{total} = \Delta S_{surr} + \Delta S_{reac.}$
$$= (-158.0 + 334.6) \text{ J K}^{-1}$$
$$= +176.6 \text{ J K}^{-1}$$

The total entropy change is positive, in accord with the requirements of the second law.

From the inequality 7.10, the quantity $(\Delta H - T\Delta S)$ must be negative for a reaction to take place (at constant T and p). ΔH is positive for an endothermic reaction, so the term $T\Delta S$ must be both positive and large enough to outweigh this contribution. Since T is positive by definition, ΔS must also be positive, as in the example above.

QUESTION 8.1 (Learning Outcome 6)

The electrical work done on the system is given by

$$w_{el} = \text{power} \times \text{time}$$
$$= (100 \text{ W}) \times (60 + 44) \text{ s}$$
$$= (100 \text{ J s}^{-1}) \times (104 \text{ s})$$
$$= 10.4 \times 10^3 \text{ J}$$
$$= 10.4 \text{ kJ}$$

For a change at constant pressure

$$\Delta H = q + w_{el} \qquad (8.1)$$

Assuming that the system is perfectly insulated, $q = 0$. So

$$\Delta H = w_{el} = 10.4\,kJ$$

Notice that this value of ΔH actually refers to the system as whole, including the container, heating coil and thermometer, as well as the water. In any accurate experiment to determine the enthalpy change for the water *alone,* allowance would have to be made for the other components.

QUESTION 8.2 *(Learning Outcomes 7 and 8)*

(a) For a change at constant pressure

$$\Delta H = q + w_{el}$$

Assuming perfect insulation, $q = 0$, so $\Delta H = w_{el}$. The electrical work done is given by

$$w_{el} = (50\,J\,s^{-1}) \times (81\,s)$$

Thus, for vaporization of 1.62 g of water,

$$\Delta H = 4\,050\,J$$

(b) ΔH for a process like this depends on the amount of water vaporized. One mole of H_2O has a mass of $(2 \times 1.007\,9 + 15.999\,4)$ g or 18.02 g. Thus, 1.62 g amounts to $(1.62\,g/18.02\,g)$ mol or 0.089 9 mol.

Then ΔH for 1 mol is given by

$$\Delta H = \frac{4\,050\,J}{0.0899\,mol}$$
$$= 45.1 \times 10^3\,J\,mol^{-1}$$
$$= 45.1\,kJ\,mol^{-1}$$

As you will see in Section 8.3, this value of ΔH is generally called the molar enthalpy of vaporization, denoted by ΔH_m. The generally accepted value for water is $\Delta H_m = 44.3\,kJ\,mol^{-1}$: the value obtained above is too high because we neglected heat loss from the system.

QUESTION 8.3 *(Learning Outcome 8)*

To calculate the molar enthalpy change for the reaction

$$H^+(aq) + OH^-(aq) = H_2O(l) \qquad (8.5)$$

you must assume that all of the acid and base reacted (they were present initially in equal amounts). Now, 0.1 litre of 2.0 mol litre^{-1} acid were used, which contained

$$2.0\,mol\,litre^{-1} \times 0.1\,litre = 0.2\,mol\,of\,H^+(aq)$$

If all the acid is consumed, then the 'amount of reaction' is 0.2 mol. From the text, $\Delta H = -11.0\,kJ$. So

$$\Delta H_m = -\frac{11.0\,kJ}{0.2\,mol} = -55.0\,kJ\,mol^{-1}$$

Notice that exactly the same result would be obtained by considering the reaction of all the base (OH^-, aq) initially present.

QUESTION 8.4 (Learning Outcome 9)

Since $\Delta H = -149.5\,kJ$ when 0.50 mol of lithium reacts completely with oxygen to give 0.25 mol of Li_2O, we can write:

$$\tfrac{1}{2}Li(s) + \tfrac{1}{8}O_2(g) = \tfrac{1}{4}Li_2O(s); \Delta H_m = -149.5\,kJ\,mol^{-1} \qquad (Q.8)$$

where ΔH_m in this case refers to the balanced equation containing half a mole of lithium.

The values of ΔH_m required can then be obtained by multiplying this balanced equation, and its ΔH_m value, by four, eight and sixteen, respectively:

$$2Li(s) + \tfrac{1}{2}O_2(g) = Li_2O(s); \quad \Delta H_m = -598.0\,kJ\,mol^{-1} \qquad (Q.9)$$

$$4Li(s) + O_2(g) = 2Li_2O(s); \Delta H_m = -1\,196.0\,kJ\,mol^{-1} \qquad (Q.10)$$

$$8Li(s) + 2O_2(g) = 4Li_2O(s); \Delta H_m = -2\,392.0\,kJ\,mol^{-1} \qquad (Q.11)$$

QUESTION 8.5 (Learning Outcomes 7, 8 and 9)

This question is similar to the example discussed in the text: again there are two parts to the experiment, both conducted at constant pressure in an insulated container.

When the reaction takes place under these conditions, the corresponding enthalpy change is given by

$$\Delta H_1 = H(\text{products, 299.5 K}) - H(\text{reactants, 298.15 K})$$

In this part of the experiment, no electrical work is done ($w_1 = 0$) and, *assuming* perfect insulation, no heat is transferred ($q_1 = 0$). So

$$\Delta H_1 = q_1 + w_1$$
$$= 0$$

The calibration experiment corresponds to the enthalpy change

$$\Delta H_2 = H(\text{products, 299.5 K}) - H(\text{products, 298.15 K})$$

In this part of the experiment, electrical work is done on the system, so

$$w_2 = w_{el}$$
$$= (50\,J\,s^{-1}) \times (61.5\,s)$$
$$= 3\,075\,J$$

Again, the calorimeter must be assumed to be perfectly insulated, so $q_2 = 0$ and

$$\Delta H_2 = q_2 + w_2$$
$$= 3\,075\,J$$

Then the enthalpy change for the reaction at 298.15 K is given by

$$\Delta H = H(\text{products, 298.15 K}) - H(\text{reactants, 298.15 K})$$
$$= \Delta H_1 - \Delta H_2$$
$$= -3\,075\,J$$

(i)   One mole of zinc has a mass of 65.38 g, so *assuming* all of the zinc is consumed, the amount of reaction is (1.308/65.38) mol. For the reaction

$$Zn(s) + 2H^+(aq) = Zn^{2+}(aq) + H_2(g)$$

$$\Delta H_m = -\frac{3\,075\,J}{(1.308/65.38)\,mol}$$

$$= -\frac{3\,075 \times 65.38}{1.308}\,J\,mol^{-1}$$

$$= -153.7 \times 10^3\,J\,mol^{-1}$$

$$= -153.7\,kJ\,mol^{-1}$$

(ii) The equation

$$\tfrac{1}{2}Zn(s) + H^+(aq) = \tfrac{1}{2}Zn^{2+}(aq) + \tfrac{1}{2}H_2(g)$$

is the result of dividing the equation in part (i) of this question by two. The value of ΔH_m must therefore also be halved, so $\Delta H_m = -76.9\,kJ\,mol^{-1}$.

QUESTION 8.6 (Learning Outcome 9)

The reaction implied by the question could be

$$Al(s) + \tfrac{3}{2}Cl_2(g) = AlCl_3(s) \tag{Q.12}$$

or

$$2Al(s) + 3Cl_2(g) = 2AlCl_3(s) \tag{Q.13}$$

or any multiple or sub-multiple of these expressions. Unless an equation is specified, a molar enthalpy change cannot be calculated. The question gives no indication of the equation referred to.

QUESTION 9.1 (Learning Outcomes 13 and 14)

From Table 9.1, the $\Delta H_f^{\ominus}$ values are $-1\,675.7\,kJ\,mol^{-1}$ and $-1\,206.9\,kJ\,mol^{-1}$, respectively. They are the $\Delta H_m^{\ominus}$ values of the reactions in which one mole of $Al_2O_3(s)$ and $CaCO_3(s)$ are formed from their elements in the standard reference states:

$$2Al(s) + \tfrac{3}{2}O_2(g) = Al_2O_3(s); \quad \Delta H_m^{\ominus} = -1\,675.7\,kJ\,mol^{-1} \tag{Q.14}$$

$$Ca(s) + C(graphite) + \tfrac{3}{2}O_2(g) = CaCO_3(s); \quad \Delta H_m^{\ominus} = -1\,206.9\,kJ\,mol^{-1} \tag{Q.15}$$

Notice that graphite rather than diamond is indicated in the $CaCO_3$ equation. This is because graphite and not diamond is the standard reference state of carbon, a fact apparent from its zero $\Delta H_f^{\ominus}$ value in Table 9.1.

QUESTION 9.2 (Learning Outcomes 13 and 14)

(i)   For the reaction

$$Na(s) + \tfrac{1}{2}Cl_2(g) = NaCl(s) \tag{5.2}$$

the *Data Book* tells us that

$$\Delta H_f^{\ominus}(NaCl, s, 298.15\,K) = -411.2\,kJ\,mol^{-1}$$

and as Equation 5.2 represents the reaction in which sodium chloride is formed from its elements, the required value of $\Delta H_m^{\ominus}$ is simply $-411.2\,kJ\,mol^{-1}$ (cf. the value given in Question 6.1). As $Na(s)$ and $Cl_2(g)$ are the reference states for these two elements, their standard enthalpies of formation are zero by definition.

(ii) For the reaction

$$PbO(s) + C(graphite) = Pb(s) + CO(g) \tag{3.1}$$

from the discussion in the text, summarized by Equation 9.18, we have

$$\Delta H_m^{\ominus} = \Delta H_f^{\ominus}(Pb, s) + \Delta H_f^{\ominus}(CO, g) - \Delta H_f^{\ominus}(PbO, s) - \Delta H_f^{\ominus}(C, graphite)$$

Again, the enthalpies of formation of the elements in their reference states, $Pb(s)$ and $C(graphite)$, are zero by definition. Using values from the *Data Book*,

$$\Delta H_m^{\ominus} = \{0 + (-110.5) - (-219.0) - 0\} \text{ kJ mol}^{-1}$$
$$= +108.5 \text{ kJ mol}^{-1}$$

(iii) Looking at the values for the equation

$$SOCl_2(l) + H_2O(l) = SO_2(g) + 2HCl(g) \tag{6.4}$$

we find that none of the entries is zero, and that $\Delta H_f^{\ominus}$ for HCl must be multiplied by two. Hence

$$\Delta H_m^{\ominus} = \Delta H_f^{\ominus}(SO_2, g) + 2\,\Delta H_f^{\ominus}(HCl, g) - \Delta H_f^{\ominus}(SOCl_2, l) - \Delta H_f^{\ominus}(H_2O, l)$$
$$= \{(-296.8) + 2 \times (-92.3) - (-242.7) - (-285.8)\} \text{ kJ mol}^{-1}$$
$$= +47.1 \text{ kJ mol}^{-1} \text{ (cf. the value given in Section 6)}$$

Notice that a different value would be obtained if gaseous water were specified.

(iv) For the reaction

$$3K(s) + AlCl_3(s) = 3KCl(s) + Al(s) \tag{3.9}$$

the entries for $K(s)$ and $Al(s)$ are zero, so

$$\Delta H_m^{\ominus} = 3\,\Delta H_f^{\ominus}(KCl, s) - \Delta H_f^{\ominus}(AlCl_3, s)$$
$$= \{3 \times (-436.7) - (-704.2)\} \text{ kJ mol}^{-1}$$
$$= -605.9 \text{ kJ mol}^{-1}$$

Notice that the values in parts (i) and (iii) are those quoted in Section 6: they were *standard molar* enthalpy changes.

QUESTION 9.3 (Learning Outcome 14)

(i) For the equation

$$Zn(s) + 2H^+(aq) = Zn^{2+}(aq) + H_2(g) \tag{2.4}$$

all the values of $\Delta H_f^{\ominus}$ are zero by definition except that of $Zn^{2+}(aq)$, so

$$\Delta H_m^{\ominus} = \Delta H_f^{\ominus}(Zn^{2+}, aq)$$
$$= -153.9 \text{ kJ mol}^{-1}$$

The important implications of this result are taken up in Section 11.2.

(ii) For the equation

$$Mg(s) + 2Ag^+(aq) = Mg^{2+}(aq) + 2Ag(s) \tag{2.13}$$

the entries in the $\Delta H_f^{\ominus}$ column are zero for the elements, $Mg(s)$ and $Ag(s)$, so

$$\Delta H_m^{\ominus} = \Delta H_f^{\ominus}(Mg^{2+}, aq) - 2\,\Delta H_f^{\ominus}(Ag^+, aq)$$
$$= \{(-466.9) - 2 \times (105.6)\} \text{ kJ mol}^{-1}$$
$$= -678.1 \text{ kJ mol}^{-1}$$

(iii) For the equation

$$H^+(aq) + OH^-(aq) = H_2O(l) \qquad (8.5)$$

the entry for $H^+(aq)$ is zero by definition. Thus

$$\Delta H_m^{\ominus} = \Delta H_f^{\ominus}(H_2O, l) - \Delta H_f^{\ominus}(OH^-, aq)$$
$$= \{(-285.8) - (-230.0)\} \text{ kJ mol}^{-1}$$
$$= -55.8 \text{ kJ mol}^{-1} \text{ (cf. the value calculated in Question 8.3)}$$

QUESTION 9.4 (Learning Outcome 14)

The reactions in the question are:

$$Mg(s) + Cl_2(g) = MgCl_2(s); \ \Delta H_m^{\ominus} = -641.3 \text{ kJ mol}^{-1} \qquad (9.27)$$
$$MgCl_2(s) = Mg^{2+}(aq) + 2Cl^-(aq); \ \Delta H_m^{\ominus} = -160.0 \text{ kJ mol}^{-1} \qquad (Q.16)$$

So, for the *sum* of these two reactions,

$$Mg(s) + Cl_2(g) = Mg^{2+}(aq) + 2Cl^-(aq) \qquad (Q.17)$$
$$\Delta H_m^{\ominus} = \{-641.3 + (-160.0)\} \text{ kJ mol}^{-1}$$
$$= -801.3 \text{ kJ mol}^{-1}$$

Now, $\Delta H_f^{\ominus}(Cl^-, aq, 298.15 \text{ K}) = -167.2 \text{ kJ mol}^{-1}$. So

$$\Delta H_f^{\ominus}(Mg^{2+}, aq, 298.15 \text{ K}) = \{-801.3 - 2 \times (-167.2)\} \text{ kJ mol}^{-1}$$
$$= -466.9 \text{ kJ mol}^{-1}$$

This is the entry against $Mg^{2+}(aq)$ in the *Data Book*.

QUESTION 10.1 (Learning Outcome 10)

(a) From a generalization of Equation 10.3:

$$\Delta S_{fus} = \frac{\Delta H_{fus}}{T_{fus}} = \frac{6.4 \times 10^3 \text{ J mol}^{-1}}{172 \text{ K}}$$
$$= 37.2 \text{ J K}^{-1} \text{ mol}^{-1}$$

(b) Again from Equation 10.3:

$$\Delta S_{vap} = \frac{\Delta H_{vap}}{T_{vap}} = \frac{20.4 \times 10^3 \text{ J mol}^{-1}}{239 \text{ K}}$$
$$= 85.4 \text{ J K}^{-1} \text{ mol}^{-1}$$

Notice that the positive values of ΔS_m for fusion, vaporization (and *sublimation*, solid $\rightarrow$ gas) are an inevitable consequence of the fact that these processes are endothermic.

(c) The answer to part (b) is the *molar* entropy of vaporization. One mole of Cl_2 has a mass of 70.9 g, so 7.09 g represents 0.100 mol. Thus,

$$\Delta S = (85.4 \text{ J K}^{-1} \text{ mol}^{-1}) \times (0.100 \text{ mol})$$
$$= 8.54 \text{ J K}^{-1}$$

QUESTION 10.2 (Learning Outcome 11)

(a) The increment covers the temperature range from 20.64 K to 25.22 K, for which the mean temperature, T', is 22.93 K. Then,

$$\Delta S' = \frac{q}{T'} = \frac{w_{el}}{T'} = \frac{47.275\,J}{22.93\,K} = 2.062\ J\,K^{-1}$$

(b) Equation 10.5 now tells us that we can obtain an approximate value of the entropy difference between one mole of chlorine at 90.00 K and at 13.50 K by adding up the values of $\Delta S'$ in column 6 of Table 10.2. The result is 39.388 J K^{-1}.

QUESTION 10.3 (Learning Outcome 11)

The entropy change is the area beneath the curve between 0 K and 90 K, and this is the sum of areas a, b, c, d and e.

area of rectangle a = (90 − 14) K × (0.24 J K^{-2})

$\qquad\qquad$ = 76 × 0.24 J K^{-1}

$\qquad\qquad$ = 18.24 J K^{-1}

area of rectangle b = (90 − 23) K × (0.45 − 0.24) J K^{-2}

$\qquad\qquad$ = 67 × 0.21 J K^{-1}

$\qquad\qquad$ = 14.07 J K^{-1}

area of triangle $c\quad = \frac{1}{2}$ base × height

$\qquad\qquad$ = 0.5 × (14 K) × (0.24 J K^{-2})

$\qquad\qquad$ = 1.68 J K^{-1}

area of triangle $d\quad$ = 0.5 × (23 − 14) K × (0.45 − 0.24) J K^{-2}

$\qquad\qquad$ = 0.5 × 9 × 0.21 J K^{-1}

$\qquad\qquad$ = 0.95 J K^{-1}

estimated area of 'triangle' e = 0.5 × (90 − 23) K × (0.61 − 0.45) J K^{-2}

$\qquad\qquad$ = 0.5 × 67 × 0.16 J K^{-1}

$\qquad\qquad$ = 5.36 J K^{-1}

So $\quad\Delta S$ = (18.24 + 14.07 + 1.68 + 0.95 + 5.36) J K^{-1}

$\qquad$ = 40.3 J K^{-1}

The results in Figure 10.4 refer to 1 mol of chlorine, so the molar entropy increment is ΔS_m = 40.3 J K^{-1} mol^{-1}.

Clearly, this method is approximate: the area e, for example, deviates considerably from a triangle. A more accurate but tedious approach is to count the little squares beneath the curve, including those that are more than half below, and rejecting those that are more than half above. This gives a value of about 40.8 J K^{-1} mol^{-1}.

QUESTION 10.4 (Learning Outcome 14)

(a)$\quad$(i)$\quad$Na(s) + $\frac{1}{2}$Cl$_2$(g) = NaCl(s)$\hfill$(5.2)

By Equation 10.11, and taking values from the Data Book:

$\quad\Delta S_m^{\ominus}$ = $S^{\ominus}$(NaCl, s) − $S^{\ominus}$(Na, s) − $\frac{1}{2}S^{\ominus}$(Cl$_2$, g)

$\qquad$ = {72.1 − (51.2) − $\frac{1}{2}$(223.1)} J K^{-1} mol^{-1}

$\qquad$ = −90.7 J K^{-1} mol^{-1} (cf. the value quoted in Section 7.3)

Notice two points. Firstly, the $S^{\ominus}$ values for elements are *not* zero: absolute entropies are listed in the Data Book, not entropies of formation. Secondly, the subscript m on $\Delta S_m^{\ominus}$ now refers specifically to the equation as written in the question. If the equation were written:

$$2Na(s) + Cl_2(g) = 2NaCl(s) \tag{Q.18}$$

then $\quad \Delta S_m^{\ominus} = -181.4\,J\,K^{-1}\,mol^{-1}$

If you got this question right and you do not have time to try all the remaining examples, do one from part (b) and then look at the values of $\Delta S_m^{\ominus}$ below, before returning to the main text.

(ii) $Mg(s) + \frac{1}{2}O_2(g) = MgO(s)$ (2.5)

$$\Delta S_m^{\ominus} = S^{\ominus}(MgO, s) - S^{\ominus}(Mg, s) - \frac{1}{2}S^{\ominus}(O_2, g)$$
$$= \{26.9 - 32.7 - \frac{1}{2}(205.1)\}\,J\,K^{-1}\,mol^{-1}$$
$$= -108.4\,J\,K^{-1}\,mol^{-1}$$

(iii) $CaCO_3(s) = CaO(s) + CO_2(g)$ (9.10)

$$\Delta S_m^{\ominus} = S^{\ominus}(CaO, s) + S^{\ominus}(CO_2, g) - S^{\ominus}(CaCO_3, s)$$
$$= \{39.7 + 213.7 - (92.9)\}\,J\,K^{-1}\,mol^{-1}$$
$$= 160.5\,J\,K^{-1}\,mol^{-1}$$

(iv) $N_2O_4(g) = 2NO_2(g)$ (9.19)

$$\Delta S_m^{\ominus} = 2S^{\ominus}(NO_2, g) - S^{\ominus}(N_2O_4, g)$$
$$= \{2 \times (240.1) - 304.3\}\,J\,K^{-1}\,mol^{-1}$$
$$= 175.9\,J\,K^{-1}\,mol^{-1}$$

(b) (v) $Zn(s) + 2H^+(aq) = Zn^{2+}(aq) + H_2(g)$ (2.4)

$$\Delta S_m^{\ominus} = S^{\ominus}(Zn^{2+}, aq) + S^{\ominus}(H_2, g) - S^{\ominus}(Zn, s) - 2S^{\ominus}(H^+, aq)$$
$$= (-112.1 + 130.7 - 41.6 - 0)\,J\,K^{-1}\,mol^{-1}$$
$$= -23.0\,J\,K^{-1}\,mol^{-1}$$

(vi) $Cu(s) + 2H^+(aq) = Cu^{2+}(aq) + H_2(g)$ (2.9)

$$\Delta S_m^{\ominus} = S^{\ominus}(Cu^{2+}, aq) + S^{\ominus}(H_2, g) - S^{\ominus}(Cu, s) - 2S^{\ominus}(H^+, aq)$$
$$= (-99.6 + 130.7 - 33.2 - 0)\,J\,K^{-1}\,mol^{-1}$$
$$= -2.1\,J\,K^{-1}\,mol^{-1}$$

QUESTION 11.1 *(Learning Outcomes 12, 14 and 15)*

(a) $SOCl_2(l) + H_2O(l) = SO_2(g) + 2HCl(g)$ (6.4)

$$\Delta S_m^{\ominus} = S^{\ominus}(SO_2, g) + 2S^{\ominus}(HCl, g) - S^{\ominus}(SOCl_2, l) - S^{\ominus}(H_2O, l)$$
$$= \{248.2 + (2 \times 186.9) - 217.5 - 69.9\}\,J\,K^{-1}\,mol^{-1}$$
$$= 334.6\,J\,K^{-1}\,mol^{-1}$$

(b) $\Delta G_m^{\ominus} = \Delta H_m^{\ominus} - T\Delta S_m^{\ominus}$ (11.4)

$$= 47.1\,kJ\,mol^{-1} - 298.15\,K \times (334.6\,J\,K^{-1}\,mol^{-1})$$
$$= 47.1\,kJ\,mol^{-1} - 99\,761\,J\,mol^{-1}$$
$$= (47.1 - 99.8)\,kJ\,mol^{-1}$$
$$= -52.7\,kJ\,mol^{-1}$$

(c)   For Reaction 6.4, $\Delta H_m^{\ominus}$ is positive so, as noted on p. 44, the occurrence of the reaction violates Thomsen's hypothesis. However, because $\Delta S_m^{\ominus}$ is large and positive, our calculated value of $\Delta G_m^{\ominus}$ turns out to be negative. Thus, the essential requirement for a thermodynamically favourable reaction, $\Delta G_m^{\ominus} < 0$ is fulfilled.

QUESTION 11.2 (Learning Outcome 1)

From Equation 11.7, Equation 11.8 or Table 11.1, if $\Delta G_m^\ominus = 0$, then $K = 1$. The equilibrium constant for the reaction A = B has the form

$$K = \frac{[B]}{[A]}$$

If $K = 1$, then the concentrations of A and B at equilibrium must be the same. Since B is formed from A, this must mean that a half or 50% of the reactant remains.

This question highlights an extremely important point. $\Delta G_m^\ominus = 0$ does *not* mean that the reaction cannot happen: it simply does not go to completion.

QUESTION 11.3 (Learning Outcomes 12, 14 and 15)

(i) $Na(s) + \frac{1}{2}Cl_2(g) = NaCl(s)$ (5.2)

$\Delta H_m^\ominus = -411.2 \, \text{kJ mol}^{-1}$ (Question 9.2)

$\Delta S_m^\ominus = -90.7 \, \text{J K}^{-1} \, \text{mol}^{-1}$ (Question 10.4)

$\Delta G_m^\ominus = -411.2 \, \text{kJ mol}^{-1} - (298.15 \, \text{K}) \times (-90.7 \, \text{J K}^{-1} \, \text{mol}^{-1})$

$= (-411.2 + 27.0) \, \text{kJ mol}^{-1}$

$= -384.2 \, \text{kJ mol}^{-1}$

According to Table 11.1, this large negative value suggests that the reaction should go effectively to completion under ambient conditions. Conversely, the reverse reaction should be undetectable in any practical sense. These results confirm the qualitative insights developed in Section 5.

(ii) $Cu(s) + 2H^+(aq) = Cu^{2+}(aq) + H_2(g)$ (2.9)

All values of $\Delta H_m^\ominus$ are zero by definition, save that for $Cu^{2+}(aq)$, so,

$\Delta H_m^\ominus = \Delta H_f^\ominus (Cu^{2+}, aq) = 64.8 \, \text{kJ mol}^{-1}$

$\Delta S_m^\ominus = -2.1 \, \text{J K}^{-1} \, \text{mol}^{-1}$ (Question 10.4)

$\Delta G_m^\ominus = 64.8 \, \text{kJ mol}^{-1} - (298.15 \, \text{K} \times -2.1 \, \text{J K}^{-1} \, \text{mol}^{-1})$

$= (64.8 + 0.63) \, \text{kJ mol}^{-1}$

$= +65.4 \, \text{kJ mol}^{-1}$

Now you can interpret the lack of reaction when copper is added to acid, in a precise way. Evidently the reaction is unfavourable from a thermodynamic point of view, so you would not expect anything to happen.

(iii) $Zn(s) + 2H^+(aq) = Zn^{2+}(aq) + H_2(g)$ (2.4)

$\Delta H_m^\ominus = -153.9 \, \text{kJ mol}^{-1}$ (Question 9.3)

$\Delta S_m^\ominus = -23.0 \, \text{J K}^{-1} \, \text{mol}^{-1}$ (Question 10.4)

$\Delta G_m^\ominus = -153.9 \, \text{kJ mol}^{-1} - (298.15 \, \text{K} \times -23.0 \, \text{J K}^{-1} \, \text{mol}^{-1})$

$= (-153.9 + 6.86) \, \text{kJ mol}^{-1}$

$= -147.0 \, \text{kJ mol}^{-1}$

The value is large and negative. The reaction happens in practice (as you saw in Activity 2.1) because it is *both* thermodynamically favourable *and* has a reasonable *rate* of reaction. This point is examined more fully in Section 13.

QUESTION 11.4 (Learning Outcome 14)

Using Equation 11.12:

$$\Delta G_m^{\ominus} = \Delta G_f^{\ominus}(SO_2, g) + 2\,\Delta G_f^{\ominus}(HCl, g) - \Delta G_f^{\ominus}(SOCl_2, l) - \Delta G_f^{\ominus}(H_2O, l)$$

$$= \{-300.2 + 2 \times (-95.3) - (-201.0) - (-237.1)\}\ \text{kJ mol}^{-1}$$

$$= -52.7\ \text{kJ mol}^{-1}$$

This is exactly the value obtained from $\Delta H_m^{\ominus}$ and $\Delta S_m^{\ominus}$ in Question 11.1.

QUESTION 11.5 (Learning Outcomes 13 and 14)

For the aqueous ion $Sc^{3+}(aq)$, the formation reaction is

$$Sc(s) + 3H^+(aq) = Sc^{3+}(aq) + \tfrac{3}{2}H_2(g) \qquad\qquad (Q.19)$$

so

$$\Delta S_f^{\ominus} = S^{\ominus}(Sc^{3+}, aq) + \tfrac{3}{2}S^{\ominus}(H_2, g) - S^{\ominus}(Sc, s) - 3S^{\ominus}(H^+, aq)$$

$$= \{-255.2 + \tfrac{3}{2}(130.7) - 34.6 - 0\}\ \text{J K}^{-1}\,\text{mol}^{-1}$$

$$= -93.8\ \text{J K}^{-1}\,\text{mol}^{-1}$$

$$\Delta G_f^{\ominus} = \Delta H_f^{\ominus} - T\,\Delta S_f^{\ominus} \qquad\qquad (11.9)$$

so

$$\Delta H_f^{\ominus} = \Delta G_f^{\ominus} + T\,\Delta S_f^{\ominus}$$

$$= (-586.6\ \text{kJ mol}^{-1}) + (298.15\ \text{K}) \times (-93.8\ \text{J K}^{-1}\,\text{mol}^{-1})$$

$$= (-586.6 - 28.0)\ \text{kJ mol}^{-1}$$

$$= -614.6\ \text{kJ mol}^{-1}$$

For the oxide, the appropriate formation reaction is

$$2Sc(s) + \tfrac{3}{2}O_2(g) = Sc_2O_3(s) \qquad\qquad (Q.20)$$

So

$$\Delta S_f^{\ominus} = S^{\ominus}(Sc_2O_3, s) - 2S^{\ominus}(Sc, s) - \tfrac{3}{2}S^{\ominus}(O_2, g)$$

$$= \{(+77.0) - 2(34.6) - \tfrac{3}{2}(205.1)\}\ \text{J K}^{-1}\,\text{mol}^{-1}$$

$$= -299.9\ \text{J K}^{-1}\,\text{mol}^{-1}$$

Then

$$\Delta G_f^{\ominus} = \Delta H_f^{\ominus} - T\,\Delta S_f^{\ominus}$$

$$= (-1\,908.8\ \text{kJ mol}^{-1}) - (298.15\ \text{K}) \times (-299.9\ \text{J K}^{-1}\,\text{mol}^{-1})$$

$$= (-1\,908.8 + 89.4)\ \text{kJ mol}^{-1}$$

$$= -1\,819.4\ \text{kJ mol}^{-1}$$

QUESTION 12.1 (Learning Outcomes 9, 12 and 14)

(i) $-150.6\ \text{kJ mol}^{-1}$; (ii) $520.4\ \text{kJ mol}^{-1}$; (iii) $-233.2\ \text{kJ mol}^{-1}$; (iv) $1\,463.4\ \text{kJ mol}^{-1}$.

(i) Subtract the silver equation and its $\Delta G_m^{\ominus}$ value from the zinc equation and its $\Delta G_m^{\ominus}$ value.

(ii) Subtract the magnesium line from the copper line, and then multiply by two.

(iii) Subtract the silver line from the iron line, and then multiply by two.

(iv) Subtract the aluminium line from the mercury line, and then multiply by six.

QUESTION 12.2 (Learning Outcomes 13 and 14)

The completed entries are in Table Q.1.

Table Q.1 Thermodynamic data at 298.15 K for $La^{3+}(aq)$ and $Pu^{3+}(aq)$

Substance	State	$\dfrac{\Delta H_f^{\ominus}}{kJ\ mol^{-1}}$	$\dfrac{\Delta G_f^{\ominus}}{kJ\ mol^{-1}}$	$\dfrac{S^{\ominus}}{J\ K^{-1}\ mol^{-1}}$
La^{3+}	aq	−709.0	−683.9	−223.7
Pu^{3+}	aq	−593.0	−585.3	−171.7

(i) For the two ions, the $\Delta H_f^{\ominus}$ values are the values of $\Delta H_m^{\ominus}$ for the reactions

$$La(s) + 3H^+(aq) = La^{3+}(aq) + \tfrac{3}{2}H_2(g) \tag{12.5}$$
$$Pu(s) + 3H^+(aq) = Pu^{3+}(aq) + \tfrac{3}{2}H_2(g) \tag{12.6}$$

This can be shown for the lanthanum reaction in the following way:

$$\Delta H_m^{\ominus} = \Sigma \Delta H_f^{\ominus}(products) - \Sigma \Delta H_f^{\ominus}(reactants) \tag{9.18}$$
$$= \Delta H_f^{\ominus}(La^{3+}, aq) + \tfrac{3}{2}\Delta H_f^{\ominus}(H_2, g) - \Delta H_f^{\ominus}(La, s) - 3\Delta H_f^{\ominus}(H^+, aq)$$
$$= \Delta H_f^{\ominus}(La^{3+}, aq) + 0 - 0 - 0$$

As $\Delta H_m^{\ominus} = -709.0\ kJ\ mol^{-1}$

$$\Delta H_f^{\ominus}(La^{3+}, aq) = -709.0\ kJ\ mol^{-1}$$

Similarly,

$$\Delta H_f^{\ominus}(Pu^{3+}, aq) = -593.0\ kJ\ mol^{-1}$$

(ii) Likewise, the $\Delta G_f^{\ominus}$ values for the two ions are the values of $\Delta G_m^{\ominus}$ for the reactions

$$La(s) + 3H^+(aq) = La^{3+}(aq) + \tfrac{3}{2}H_2(g) \tag{12.5}$$
$$Pu(s) + 3H^+(aq) = Pu^{3+}(aq) + \tfrac{3}{2}H_2(g) \tag{12.6}$$

This can be shown for the lanthanum reaction in the following way:

$$\Delta G_m^{\ominus} = \Sigma \Delta G_f^{\ominus}(products) - \Sigma \Delta G_f^{\ominus}(reactants) \tag{11.12}$$
$$= \Delta G_f^{\ominus}(La^{3+}, aq) + \tfrac{3}{2}\Delta G_f^{\ominus}(H_2, g) - \Delta G_f^{\ominus}(La, s) - 3\Delta G_f^{\ominus}(H^+, aq)$$
$$= \Delta G_f^{\ominus}(La^{3+}, aq) + 0 - 0 - 0$$

For the lanthanum reaction

$$\Delta H_m^{\ominus} = -709.0\ kJ\ mol^{-1} \text{ and } \Delta S_m^{\ominus} = -84.1\ J\ K^{-1}\ mol^{-1}$$
$$\Delta G_m^{\ominus} = \Delta H_m^{\ominus} - T\Delta S_m^{\ominus}$$
$$= -709.0\ kJ\ mol^{-1} - (298.15 \times -84.1)\ J\ mol^{-1}$$
$$= -709.0 + 25.1\ kJ\ mol^{-1}$$
$$= -683.9\ kJ\ mol^{-1}$$

For the plutonium reaction

$$\Delta H_m^{\ominus} = -593.0\ kJ\ mol^{-1} \text{ and } \Delta S_m^{\ominus} = -25.9\ J\ K^{-1}\ mol^{-1}$$
$$\Delta G_m^{\ominus} = -593.0\ kJ\ mol^{-1} - (298.15 \times -25.9)\ J\ mol^{-1}$$
$$= -593.0 + 7.7\ kJ\ mol^{-1}$$
$$= -585.3\ kJ\ mol^{-1}$$

(iii) For the lanthanum reaction

$\Delta S_m^{\ominus} = -84.1 \, \text{J K}^{-1} \, \text{mol}^{-1}$

$= S^{\ominus}(\text{La}^{3+}, \text{aq}) + \frac{3}{2}S^{\ominus}(\text{H}_2, \text{g}) - S^{\ominus}(\text{La}, \text{s}) - 3S^{\ominus}(\text{H}^+, \text{aq})$

$= S^{\ominus}(\text{La}^{3+}, \text{aq}) + (196.1 - 56.5 - 0) \, \text{J K}^{-1} \, \text{mol}^{-1}$

$S^{\ominus}(\text{La}^{3+}, \text{aq}) = (-84.1 - 196.1 + 56.5) \, \text{J K}^{-1} \, \text{mol}^{-1}$

$= -223.7 \, \text{J K}^{-1} \, \text{mol}^{-1}$

For the plutonium reaction

$S^{\ominus}(\text{Pu}^{3+}, \text{aq}) = (-25.9 - 196.1 + 50.3) \, \text{J K}^{-1} \, \text{mol}^{-1}$

$= -171.7 \, \text{J K}^{-1} \, \text{mol}^{-1}$

QUESTION 12.3 (Learning Outcome 14)

According to the calculations in Question 12.2:

$$\text{La(s)} + 3\text{H}^+(\text{aq}) = \text{La}^{3+}(\text{aq}) + \tfrac{3}{2}\text{H}_2(\text{g}); \ \Delta G_m^{\ominus} = -683.9 \, \text{kJ mol}^{-1} \qquad (12.5)$$

$$\text{Pu(s)} + 3\text{H}^+(\text{aq}) = \text{Pu}^{3+}(\text{aq}) + \tfrac{3}{2}\text{H}_2(\text{g}); \ \Delta G_m^{\ominus} = -585.3 \, \text{kJ mol}^{-1} \qquad (12.6)$$

Converting the equations into a form in which there is one mole of $\text{H}^+(\text{aq})$ on the left and half a mole of H_2 on the right:

$$\tfrac{1}{3}\text{La(s)} + \text{H}^+(\text{aq}) = \tfrac{1}{3}\text{La}^{3+}(\text{aq}) + \tfrac{1}{2}\text{H}_2(\text{g}); \ \Delta G_m^{\ominus} = -228.0 \, \text{kJ mol}^{-1} \qquad (Q.21)$$

$$\tfrac{1}{3}\text{Pu(s)} + \text{H}^+(\text{aq}) = \tfrac{1}{3}\text{Pu}^{3+}(\text{aq}) + \tfrac{1}{2}\text{H}_2(\text{g}); \ \Delta G_m^{\ominus} = -195.1 \, \text{kJ mol}^{-1} \qquad (Q.22)$$

These values put lanthanum between sodium and magnesium, and plutonium between magnesium and aluminium.

QUESTION 13.1 (Learning Outcome 15)

Reactions (i) and (iii) have negative values of $\Delta G_m^{\ominus}$, so in these cases, equilibrium lies to the right. Ignoring the possibility of slow rates of reaction, these reactions should occur.

QUESTION 13.2 (Learning Outcomes 14, 15 and 16)

From Table 12.1, for the first reaction

$\Delta G_m^{\ominus} = 2\{32.7 - (-227.5)\} \, \text{kJ mol}^{-1}$

$= 520.4 \, \text{kJ mol}^{-1}$

and for the second reaction

$\Delta G_m^{\ominus} = 6(-161.7 - 32.7) \, \text{kJ mol}^{-1}$

$= -1 \, 166.4 \, \text{kJ mol}^{-1}$

Thus, in the first case, the lack of reaction is due to thermodynamic stability; in the second case it must be due to kinetic stability. In the second case we can attribute the kinetic stability to the oxide coating on the aluminium.

QUESTION 13.3 (*Learning Outcomes 14, 15 and 16*)

(a) A conceivable reaction here would be

$C(\text{diamond}) + O_2(g) = CO_2(g)$

From the *Data Book* we find

$\Delta G_m^{\ominus} = \Delta G_f^{\ominus}(CO_2, g) - \Delta G_f^{\ominus}(C, \text{diamond}) - \Delta G_f^{\ominus}(O_2, g)$

$\phantom{\Delta G_m^{\ominus}} = (-394.4 - 2.9 - 0)\,\text{kJ mol}^{-1} = -397.3\,\text{kJ mol}^{-1}$

(b) A conceivable reaction would be

$Mg(s) + ZnCl_2(s) = MgCl_2(s) + Zn(s)$

From the *Data Book* we find

$\Delta G_m^{\ominus} = \Delta G_f^{\ominus}(MgCl_2, s) + \Delta G_f^{\ominus}(Zn, s) - \Delta G_f^{\ominus}(Mg, s) - \Delta G_f^{\ominus}(ZnCl_2, s)$

$\phantom{\Delta G_m^{\ominus}} = (-591.8 + 0 - 0 + 369.4)\,\text{kJ mol}^{-1}$

$\phantom{\Delta G_m^{\ominus}} = -222.4\,\text{kJ mol}^{-1}$

Thus, in both cases, there is a thermodynamically favourable reaction that the system could undergo: both systems are therefore kinetically stable.

QUESTION 15.1 (*Learning Outcome 14*)

(i) Subtract the lead reaction from the aluminium reaction in Table 15.1, and then multiply by three:

$\Delta G_m^{\ominus} = -1\,015.5\,\text{kJ mol}^{-1}$

(ii) Subtract the lithium line from the copper line:

$\Delta G_m^{\ominus} = 431.5\,\text{kJ mol}^{-1}$

QUESTION 15.2 (*Learning Outcome 14*)

The values of $\Delta G_f^{\ominus}$ for the chlorides must be converted into values corresponding to the reaction of the different metals with the same number of moles of chlorine, $Cl_2(g)$. In Table Q.2 the values of $\Delta G_m^{\ominus}$ corresponding to half a mole of chlorine have been used. Thus, for sodium, $\Delta G_m^{\ominus}$ is equal to $\Delta G_f^{\ominus}(NaCl, s)$, but for aluminium, $\Delta G_m^{\ominus}$ is equal to $\frac{1}{3}\Delta G_f^{\ominus}(AlCl_3, s)$.

Table Q.2 Values of $\Delta G_m^{\ominus}$ for the reactions of metals with chlorine at 298.15 K

Reaction	$\Delta G_m^{\ominus}/\text{kJ mol}^{-1}$
$Na(s) + \frac{1}{2}Cl_2(g) = NaCl(s)$	−384.2
$\frac{1}{2}Ca(s) + \frac{1}{2}Cl_2(g) = \frac{1}{2}CaCl_2(s)$	−374.1
$\frac{1}{3}Al(s) + \frac{1}{2}Cl_2(g) = \frac{1}{3}AlCl_3(s)$	−209.6
$\frac{1}{2}Zn(s) + \frac{1}{2}Cl_2(g) = \frac{1}{2}ZnCl_2(s)$	−184.7
$Ag(s) + \frac{1}{2}Cl_2(g) = AgCl(s)$	−109.8
$\frac{1}{2}Hg(l) + \frac{1}{2}Cl_2(g) = \frac{1}{2}HgCl_2(s)$	−89.3
$\frac{1}{2}Cu(s) + \frac{1}{2}Cl_2(g) = \frac{1}{2}CuCl_2(s)$	−87.9

Table Q.2 does support the conclusions reached in Section 15. There are differences from the orders in Tables 12.1 and 15.1 in fine detail, but there are also broad similarities. Thus, sodium and calcium come near the top, but copper, silver and mercury are near the bottom.

QUESTION 15.3 (Learning Outcomes 15 and 16)

Reactions obtained by subtracting equations in Table 15.1 involve the reaction between solid metals and solid oxides. Reactions between solids tend to be slow because they can occur only where the surfaces of the different solids make contact. By contrast, reactions obtained by subtracting equations in Table 12.1 involve the reaction between solid metals and aqueous ions. Fresh ions are constantly being brought up to the metal surface by the turmoil in the solution, so these reactions tend to be faster than solid–solid reactions.

QUESTION 16.1 (Learning Outcome 17)

The pair of reactions with similar values of $\Delta S_m^{\ominus}$ is the pair that are analogous in the sense defined at the end of Section 16. Pair (i) does not fit this description because, although the reactions differ in the substitution of mercury for copper and the numbers in the equations are identical, the elemental mercury is gaseous and the copper is solid. Pair (iii) is not the answer because the numbers preceding $O_2(g)$ are not identical. Pair (ii) is the correct answer: it contains two reactions with the same physical states and coefficients; they differ only in the substitution of magnesium by calcium.

QUESTION 16.2 (Learning Outcomes 17 and 18)

(a) Reaction (i); (b) reactions (iv) and (v); (c) reactions (ii) and (iii).

Reaction (i) involves a decrease in the number of moles of gas. Reactions (ii) and (iii) involve no change in the number of moles of gas. Reactions (iv) and (v) involve an increase in the number of moles of gas.

QUESTION 17.1 (Learning Outcomes 14 and 19)

Note that mercury is in the gaseous state. For the decomposition reaction at 298.151K:

$$\Delta H_m^{\ominus} = \Delta H_f^{\ominus}(Hg, g) + \tfrac{1}{2}\Delta H_f^{\ominus}(O_2, g) - \Delta H_f^{\ominus}(HgO, s)$$
$$= (61.3 + 0) - (-90.8)\,\text{kJ mol}^{-1}$$
$$= 152.1\,\text{kJ mol}^{-1}$$
$$\Delta S_m^{\ominus} = S^{\ominus}(Hg, g) + \tfrac{1}{2}S^{\ominus}(O_2, g) - S^{\ominus}(HgO, s)$$
$$= (175.0 + 102.6 - 70.3)\,\text{J K}^{-1}\,\text{mol}^{-1}$$
$$= 207.3\,\text{J K}^{-1}\,\text{mol}^{-1}$$

Now use Equation 17.3:

$$T = \frac{\Delta H_m^{\ominus}(298.15\,\text{K})}{\Delta S_m^{\ominus}(298.15\,\text{K})}$$

$$= \frac{152.1\,\text{kJ mol}^{-1}}{207.3\,\text{J K}^{-1}\,\text{mol}^{-1}} = 734\,\text{K}$$

This temperature is close to the value of about 750 K, at which the value of $\Delta G_m^{\ominus}$ for the HgO line on Figure 17.5 is zero (above the boiling temperature of mercury).

QUESTION 17.2 (Learning Outcomes 15 and 20)

In Figure 17.5, the CO line drops below the MnO line at about 1 700 K, and below the ZnO line at about 1 200 K. It is above these temperatures that carbon is thermodynamically capable of reducing the oxides. At such high temperatures we expect the reactions to be fairly fast, and reduction does in fact occur in both cases. In the zinc reaction the crossover point of the ZnO and CO plots occurs above the boiling temperature of zinc. The product is thus zinc vapour and the reaction is

$$C(s) + ZnO(s) = CO(g) + Zn(g) \qquad\qquad (Q.23)$$

For Mn at 1 700 K the reaction is similar except that the product would be *liquid* manganese.

QUESTION 17.3 (Learning Outcome 21)

Statements (i), (iii) and (v) are false; statements (ii), (iv) and (vi) are true.

(i) Those metals whose oxides have the most negative values of $\Delta G_f^{\ominus}$ are hardest to extract from oxides, so from Table 15.1, copper is easier to extract than tin.

(ii) See Section 17.2.

(iii) Only those metals whose compounds have the least negative values of $\Delta G_f^{\ominus}$ are found free in nature. The lanthanides and actinides are not in this category, as, in fact, Question 12.3 and Figure 17.7 suggest.

(iv) Calcium is near the top in Tables 15.1 and Q.2, so it should be thermodynamically capable of reducing the compounds of many other metals.

(v) Only those metals whose oxides or chlorides have the least negative values of $\Delta G_f^{\ominus}$ are obtained by heating their oxides or chlorides.

(vi) Electrolysis of fused salts is expensive. It is used only where compounds of the metals have such negative values of $\Delta G_f^{\ominus}$ that normally cheaper methods such as carbon reduction become expensive because very high temperatures are needed.

QUESTION 18.1 (Learning Outcome 22)

Step a, (vi); step b, (ix); step c, (iv); step d, (vii); step e, (x); step f, (ii).

QUESTION 18.2 (Learning Outcome 22)

From Figure 18.2,

$$\Delta H_f^{\ominus} (LiF, s) = \Delta H_{atm}^{\ominus} (Li, s) + I_1(Li) + \tfrac{1}{2}D(F{-}F) - E(F) + L(LiF, s)$$
$$-616 \, kJ \, mol^{-1} = (159 + 520 + 79 - 328) \, kJ \, mol^{-1} + L(LiF, s)$$
$$L(LiF, s) = -1\,046 \, kJ \, mol^{-1}$$

This value is considerably more negative than that for NaCl; thus more energy is released when lithium fluoride is formed from its ions than when sodium chloride is formed from its ions. The reason for this is discussed in Section 20.4.

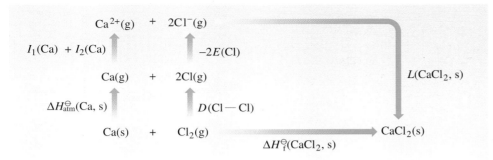

Figure Q.1
The Born–Haber cycle for calcium chloride, $CaCl_2$.

QUESTION 18.3 (Learning Outcome 22)

In the Born–Haber cycle for $CaCl_2$, which is shown in Figure Q.1, the lattice energy corresponds to the reaction

$$Ca^{2+}(g) + 2Cl^-(g) = CaCl_2(s)$$

The process

$$Ca(g) = Ca^{2+}(g) + 2e^-(g)$$

is the sum of the processes

$$Ca(g) = Ca^+(g) + e^-(g)$$
$$Ca^+(g) = Ca^{2+}(g) + e^-(g)$$

whose ionization energies are $I_1(Ca)$ and $I_2(Ca)$, respectively. The two electrons are taken up by the two chlorine atoms.

From the cycle

$$\Delta H_f^\ominus(CaCl_2, s) = \Delta H_{atm}^\ominus(Ca, s) + I_1(Ca) + I_2(Ca) + D(Cl-Cl) - 2E(Cl) + L(CaCl_2, s)$$

$$-796\,kJ\,mol^{-1} = (178 + 590 + 1\,145 + 244 - 698)\,kJ\,mol^{-1} + L(CaCl_2, s)$$

Therefore

$$L(CaCl_2, s) = -2\,255\,kJ\,mol^{-1}$$

Thus, about three times as much energy is released when one mole of solid calcium chloride is formed from gaseous Ca^{2+} and Cl^- ions as when one mole of solid sodium chloride is formed from gaseous Na^+ and Cl^- ions. If we assume that the bonding in both chlorides is ionic, then the attraction between the ions is larger in $CaCl_2$ because the solid contains doubly charged Ca^{2+} ions. Another reason for the difference is that one mole of $CaCl_2$ contains 50% more ions than one mole of NaCl, so the energy of interaction will be correspondingly greater. These ideas are discussed quantitatively in Section 20.4.

QUESTION 18.4 (Learning Outcome 21)

Those metals with low ionization energies tend to form compounds with the most negative values of $\Delta G_f^\ominus$. These metals lie near the top of series such as those in Tables 12.1 and 15.1, and are not easily extracted from their compounds. Such metals lie to the left of Figure 17.7, so one expects ionization energies to be low at the left of the Periodic Table and high at the right. This is broadly speaking correct, as a comparison of, say, rubidium and silver shows.

QUESTION 20.1 (Learning Outcome 23)

For LiF, with $n = 6$,

$$L(\text{LiF, s}) = -\frac{1.389 \times 10^5 \times 1.748 \times 1 \times 1}{201}\left(1 - \frac{1}{6}\right) \text{kJ mol}^{-1}$$

$$= -\frac{1.389 \times 10^5 \times 1.748 \times 1 \times 1 \times 5}{201 \times 6} \text{kJ mol}^{-1}$$

$$= -1\,007 \text{ kJ mol}^{-1}$$

For CaCl_2, with $n = 9$,

$$L(\text{CaCl}_2, \text{ s}) = -\frac{1.389 \times 10^5 \times 2.408 \times 2 \times 1}{274}\left(1 - \frac{1}{9}\right)\text{kJ mol}^{-1}$$

$$= -2\,170 \text{ kJ mol}^{-1}$$

The results are the same as those in the penultimate column of Table 20.2.

QUESTION 20.2 (Learning Outcomes 22 and 23)

The cycle is shown in Figure Q.2. It corresponds closely with that given for NaI in Figure 20.8. However, the standard state of bromine at room temperature is $\text{Br}_2(\text{l})$, so $\frac{1}{2}\text{Br}_2(\text{l})$ replaces $\frac{1}{2}\text{I}_2(\text{s})$ on the left-hand side of the formation reaction for the solid halide. And K, of course, replaces Na.

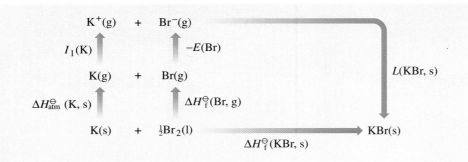

Figure Q.2
Born–Haber cycle for potassium bromide, KBr.

From the cycle,

$$\Delta H_f^{\ominus}(\text{KBr, s}) = \Delta H_{\text{atm}}^{\ominus}(\text{K, s}) + I_1(\text{K}) + \Delta H_f^{\ominus}(\text{Br, g}) - E(\text{Br}) + L(\text{KBr, s})$$

Substituting values from the *Data Book*:

$$-393.8 \text{ kJ mol}^{-1} = (89.2 + 419 + 111.9 - 324)\text{ kJ mol}^{-1} + L(\text{KBr, s})$$

$$L(\text{KBr, s}) = -690 \text{ kJ mol}^{-1} \qquad \sim 646.1$$

Note the use of $\Delta H_f^{\ominus}(\text{Br, g})$, which is greater than $\frac{1}{2}D(\text{Br}-\text{Br})$. This is because Br_2 is a liquid at 25 °C (see Section 20.5).

QUESTION 20.3 (Learning Outcome 23)
K^+ has the configuration of argon ($n = 9$) and Br^- has the configuration of krypton ($n = 10$). Thus, for KBr we use the average, $n = 9.5$:

$$L = -\frac{1.389 \times 10^5 \times 1.748 \times 1 \times 1}{330}\left(1 - \frac{1}{9.5}\right)\text{kJ mol}^{-1}$$

$$= -\frac{1.389 \times 10^5 \times 1.748 \times 0.895}{330}\,\text{kJ mol}^{-1}$$

$$= -658\,\text{kJ mol}^{-1}$$

As usual, the value is not as negative as the cycle value, but it deviates from it by only $32\,\text{kJ mol}^{-1}$, which is less than 5%.

QUESTION 20.4 (Learning Outcome 23)

Ca^{2+} has the configuration of argon ($n = 9$), and the two F^- ions in the formula unit each have the configuration of neon ($n = 7$).

The average value for the three ions is $\frac{1}{3}(9 + 7 + 7) = 7.67$.

Table 20.1 tells us that for fluorite, $M = 2.519$ so:

$$L = -\frac{1.389 \times 10^5 \times 2.519 \times 2 \times 1}{237}\left(1 - \frac{1}{7.67}\right)\text{kJ mol}^{-1}$$

$$= -\frac{1.389 \times 10^5 \times 2.519 \times 2 \times 0.870}{237}\,\text{kJ mol}^{-1}$$

$$= -2\,569\,\text{kJ mol}^{-1}$$

This value is not negative enough, but the discrepancy is only $66\,\text{kJ mol}^{-1}$, which is less than 3%.

QUESTION 20.5 (Learning Outcome 23)

In the formula unit CaF_2, there are three ions so $v = 3$. Hence

$$L(\text{CaF}_2,\ \text{s}) = -\frac{1.079 \times 10^5 \times 3 \times 2 \times 1}{(114 + 119)}\,\text{kJ mol}^{-1}$$

$$= -2\,779\,\text{kJ mol}^{-1}$$

This value is $144\,\text{kJ mol}^{-1}$ too negative. The agreement is worse than with the Born–Landé equation, the discrepancy being about 5.5%.

QUESTION 21.1 (Learning Outcome 24)

$E^{\ominus}(\text{Al}^{3+}\,|\,\text{Al}) = -1.68\,\text{V}$. First reverse the equation, and the sign of the $\Delta G_m^{\ominus}$ value:

$$\text{Al}^{3+}(\text{aq}) + \tfrac{3}{2}\text{H}_2(\text{g}) = \text{Al}(\text{s}) + 3\text{H}^+(\text{aq});\ \Delta G_m^{\ominus} = 485\,\text{kJ mol}^{-1} \qquad \text{(Q.24)}$$

Use the symbol e as shorthand for $\frac{1}{2}\text{H}_2(\text{g})$ on the side of the equation on which it appears, and for $\text{H}^+(\text{aq})$ on the other side. Since there is $\frac{3}{2}\text{H}_2(\text{g})$ on the left, and $3\text{H}^+(\text{aq})$ on the right, the equation now becomes:

$$\text{Al}^{3+}(\text{aq}) + 3\text{e} = \text{Al}(\text{s});\ \Delta G_m^{\ominus} = 485\,\text{kJ mol}^{-1} \qquad \text{(21.13)}$$

Using Equation 21.4:

$$E^{\ominus} = -\frac{485\,\text{kJ mol}^{-1}}{3 \times 96.485\,\text{kJ mol}^{-1}\,\text{V}^{-1}} = -1.68\,\text{V}$$

This value is considerably more negative than the threshold value of $-0.1\,\text{V}$, so we conclude that aluminium is a powerful reducing agent.

(handwritten annotations in right margin:)

$-2779 \times (114 + 119) = 6.474 \times 10^5$

$-\dfrac{6.47400 \times 10^5}{(114 + 119)} = -2779$

$\dfrac{-6.474 \times 10^5}{-2779} = (114 + 119) = 233$

$\left(\dfrac{-6.474 \times 10^5}{-2779}\right) - 119 = 114$

227

QUESTION 21.2 (Learning Outcome 24)

The two strongest oxidizing agents are fluorine, F_2, and chlorine, Cl_2, and the two weakest are the aqueous ions $Li^+(aq)$ and $Cs^+(aq)$. The two strongest reducing agents are lithium metal and caesium metal, and the two weakest reducing agents are the aqueous fluoride and chloride ions. The reducing agents occur on the right of the equations in Table 21.2 and the oxidizing agents on the left. The most powerful reducing agents (and therefore the least powerful oxidizing agents) are found in systems with the most negative $E^{\ominus}$ values. By contrast, the least powerful reducing agents (and therefore the most powerful oxidizing agents) are found in systems with the most positive $E^{\ominus}$ values.

QUESTION 21.3 (Learning Outcome 24)

Reactions (i), (ii), (iv) and (v) are thermodynamically favourable. The equations are:

(i) $Mg(s) + Zn^{2+}(aq) = Mg^{2+}(aq) + Zn(s)$ (Q.25)

(ii) $Zn(s) + Sn^{2+}(aq) = Zn^{2+}(aq) + Sn(s)$ (Q.26)

(iv) $Fe^{2+}(aq) + \frac{1}{2}Cl_2(g) = Fe^{3+}(aq) + Cl^-(aq)$ (Q.27)

(v) $Br^-(aq) + \frac{1}{2}Cl_2(g) = \frac{1}{2}Br_2(aq) + Cl^-(aq)$ (Q.28)

In each case, first identify the redox systems in Table 21.2. Thus, for (i) these are $Zn^{2+}|Zn$ and $Mg^{2+}|Mg$. Then identify the more negative redox potential. For (i) this is $E^{\ominus}(Mg^{2+}|Mg)$. Then the reducing agent in this system (Mg) is thermodynamically capable of reducing the oxidized state in the other system (Zn^{2+}). Hence Reaction Q.25 is favourable. The same procedure tells you that reactions (ii), (iv) and (v) in the question are also favourable. In the case of (iii), $E^{\ominus}(Ca^{2+}|Ca)$ is less than $E^{\ominus}(Zn^{2+}|Zn)$, so the favourable reaction is reduction of Zn^{2+} by calcium. This is the opposite of the change in the question, which therefore does not occur.

QUESTION 22.1 (Learning Outcome 25)

(i) Aluminium precedes sulfur in Period 3 of Figure 22.1. Consequently, sulfur has the higher ionization energy, because this quantity shows an overall increase across a Period.

(ii) Nitrogen lies above antimony in Group V of Figure 22.1. So nitrogen has the higher ionization energy because this quantity tends to decrease down a Group.

(iii) Germanium comes earlier than chlorine in its Period and lower than chlorine in its Group. The two effects of (i) and (ii) reinforce each other: chlorine has the higher ionization energy.

QUESTION 23.1 (Learning Outcome 26)

The low first and high second ionization energies of lithium and caesium mean that each atom contributes only one electron to the electron gas, so the metallic bonding is relatively weak. Moreover, as the values of $\Delta H_{atm}^{\ominus}(M, s)$ in Table 23.1 show, the bonding weakens from lithium to caesium as the radius of the ion core, and therefore the internuclear distance, increases. The rate of reaction depends on the speed with which the metal structure can be broken down. It is therefore greatest at caesium, where the metallic bonding is weakest.

QUESTION 25.1 (*Learning Outcome 26*)

Commercial sodium peroxide contains a small amount of the superoxide, NaO_2, which is formed during the combustion of sodium. The alkali metal superoxides are orange so, at the 10% level, NaO_2 confers a yellow tint on commercial Na_2O_2. In water, this impurity instantly evolves oxygen (cf. Equation 25.4), whereas the preponderant Na_2O_2 yields H_2O_2(aq) according to Equation 25.3. On boiling, the H_2O_2 formed by both Na_2O_2 and the NaO_2 impurity decomposes via Equation 25.5, generating a much larger volume of oxygen.

QUESTION 25.2 (*Learning Outcome 27*)

(a) On heating, the alkali metal hydrides decompose to the metal and hydrogen gas:

$$MH(s) = M(s) + \tfrac{1}{2} H_2(g)$$

As $\Delta H_m^{\ominus}$ for this reaction is $-\Delta H_f^{\ominus}$ (MH, s), the values in Table 25.2 show that $\Delta H_m^{\ominus}$ decreases down Group I from lithium to sodium to caesium. Since $\Delta S_m^{\ominus}$ should be similar for the three analogous reactions, the decrease in $\Delta H_m^{\ominus}$ should, according to Equation 17.3, lead to a decreasing decomposition temperature. This is consistent with the facts. As noted in the text, this change occurs because the small size of the hydride ion leads to large decreases in $-L$(MH, s) as $r(M^+)$ increases.

(b) With the iodides, the stabilities indicated by the values of $\Delta H_m^{\ominus}$ for the reaction

$$MI(s) = M(s) + \tfrac{1}{2} I_2(s)$$

increase down the Group. The iodide ion is large, so when $r(M^+)$ increases, the decreases in $-L$(MI, s) are smaller. *Compared with the hydride case*, this favours an increase in stability.

QUESTION 25.3 (*Learning Outcome 27*)

The caesium compound is the one to choose because it will be the most stable to the decomposition

$$MClF_4(s) = MF(s) + ClF_3(g) \tag{Q.29}$$

which is the reverse of the preparative reaction. This is because the ion ClF_4^- is obviously bigger than the ion F^-. Consequently, $-L$(MF, s) is larger than $-L$(MClF$_4$, s), and, in accordance with point 5 of Section 25.4, stability will increase with increasing cation size. In fact, $LiClF_4$ and $NaClF_4$ cannot be prepared, but $CsClF_4$ has been made and begins decomposing only at 300 °C.

QUESTION 26.1 (*Learning Outcome 28*)

(a) The immediate neighbours of the magnesium ion are the six oxygen atoms of six surrounding water molecules. This indicates the presence of a magnesium complex, $[Mg(H_2O)_6]^{2+}$. Figure 26.6 tries to convey the fact that the oxygen atoms of the six water ligands are octahedrally disposed around magnesium.

(b) $MgSO_4.7H_2O(s) = [Mg(H_2O)_6]^{2+}(aq) + SO_4^{2-}(aq) + H_2O(l)$ $\quad$ (Q.30)

(c) According to Table 26.1, the edta^{4-} ligand is hexadentate, so it can replace all six water molecules when a new complex is formed:

$$[Mg(H_2O)_6]^{2+}(aq) + edta^{4-}(aq) = [Mg(edta)]^{2-}(aq) + 6H_2O(l) \tag{Q.31}$$

QUESTION 27.1 (Learning Outcome 26)

The shortening of the chains suggests that the cavity within the cryptand will be reduced in size. Consequently, the match between ligand and cation size which yields the complex of greatest stability is now more likely to occur at a smaller cation radius. As only Li^+ and Na^+ have smaller radii than K^+, the different cation that forms the most stable complex with cryptand-221 must be Na^+; with cryptand-211 it must be Li^+. With cryptand-211, the stabilities decrease from lithium to caesium.

QUESTION 28.1 (Learning Outcome 26)

Compare the discussion of sodium, magnesium and aluminium in Section 23.1. In terms of the simple electron-gas model of metallic bonding, the Group II metals contain M^{2+} ion cores and two electrons per metal atom in the electron gas. This means that the ion cores are pulled together more tightly than the M^+ ion cores of the alkali metals. Thus, densities, melting temperatures and enthalpies of atomization are higher for the Group II metals.

QUESTION 28.2 (Learning Outcome 26)

If, as we surmised with the alkali metals, the rate of reaction depends on how easily the metal structure is broken down (Section 23.1), then the slower reaction of strontium and barium is attributable to their higher enthalpies of atomization.

QUESTION 28.3 (Learning Outcome 29)

There are no serious changes in the conclusions. If Ca^+ and Rb^+ are the same size, then we assume that $L(CaCl, s) = L(RbCl, s) = -692\,kJ\,mol^{-1}$. Putting this value into the third column of Table 28.4, and adding up the column of figures, $\Delta H_f^{\ominus}(CaCl, s) = -151\,kJ\,mol^{-1}$. The changes give a value of $\Delta H_f^{\ominus}(CaCl, s)$ that is $26\,kJ\,mol^{-1}$ more positive than before. Using the estimate $\Delta S_m^{\ominus} = 93.2\,J\,K^{-1}\,mol^{-1}$ for Equation 28.8 from Section 28.2, the new estimate for $\Delta H_f^{\ominus}(CaCl, s)$ gives $\Delta G_f^{\ominus}(CaCl, s) = -123\,kJ\,mol^{-1}$. Then, for the reaction

$$2CaCl(s) = Ca(s) + CaCl_2(s) \tag{28.11}$$

$$\Delta G_m^{\ominus} = -748\,kJ\,mol^{-1} - 2 \times (-123\,kJ\,mol^{-1})$$

$$= -502\,kJ\,mol^{-1}$$

As before, the calculation suggests that CaCl(s) is stable with respect to its constituent elements, but highly unstable to disproportionation.

QUESTION 28.4 (Learning Outcome 29)

The Born–Haber cycle for a difluoride, MF_2, is shown in Figure Q.3 (cf. Figure Q.1). It yields

$$\Delta H_f^{\ominus}(MF_2, s) = \Delta H_{atm}^{\ominus}(M, s) + I_1(M) + I_2(M) + D(F-F) - 2E(F) + L(MF_2, s)$$

This means that the top six figures for a fluoride in Table 28.5 must add up to give the bottom number. Thus, in the case of barium,

$$(180 + 503 + 965 + 158 - 656)\,kJ\,mol^{-1} + L(BaF_2, s) = -1\,207\,kJ\,mol^{-1}$$

whence $L(BaF_2, s) = -2\,357\,kJ\,mol^{-1}$

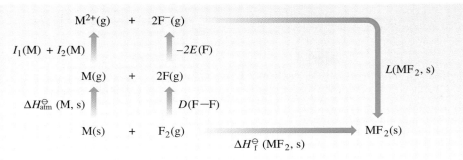

Figure Q.3
Born–Haber cycle for the difluoride MF_2.

Assuming that Ba^{2+} and Cs^{2+} have the same size, then we take $L(CsF_2, s) = L(BaF_2, s)$ $= -2\,357\,kJ\,mol^{-1}$. Then, adding up the figures in column 3 of Table 28.5,

$$\Delta H_f^{\ominus}(CsF_2, s) = (76 + 376 + 2\,236 + 158 - 656 - 2\,357)\,kJ\,mol^{-1}$$
$$= -167\,kJ\,mol^{-1}$$

Ignoring the $\Delta S_f^{\ominus}$ term, which will be fairly small, we see that this value suggests that $CsF_2(s)$ may be stable with respect to its constituent elements. Its stability is, however, much more likely to be determined by the decomposition to $CsF(s)$, a fluoride in which the cation has a noble gas configuration:

$$CsF_2(s) = CsF(s) + \tfrac{1}{2}F_2(g) \tag{Q.32}$$

Using our estimated value for $\Delta H_f^{\ominus}(CsF_2, s)$, and the value of $\Delta H_f^{\ominus}(CsF, s)$ in the *Data Book*:

$$\Delta H_m^{\ominus} = \Delta H_f^{\ominus}(CsF, s) - \Delta H_f^{\ominus}(CsF_2, s)$$
$$= -554\,kJ\,mol^{-1} - (-167\,kJ\,mol^{-1})$$
$$= -387\,kJ\,mol^{-1}$$

As Equation Q.32 occurs with an increase in the number of moles of gas, $\Delta S_m^{\ominus}$ will be positive and $\Delta G_m^{\ominus}$ even more negative than $\Delta H_m^{\ominus}$. Thus, our calculations suggest that CsF_2 will be very unstable with respect to decomposition into CsF and fluorine. It is therefore not surprising that it has not been prepared.

QUESTION 28.5 (Learning Outcome 26)

According to Equation 28.24, each P_4O_{10} molecule is converted into four PO_4^{3-} ions. Each of the P_4O_{10} molecules shown in Structure **28.1** contains six P—O—P units. In the four PO_4^{3-} ions (Structure **28.2**), these are replaced by twelve P–O⁻ bonds. Thus, two P—O bonds (in each P—O—P unit) are replaced by two P—O⁻ bonds. This parallels what happens in the reactions of CaO with CO_2 and SiO_2.

QUESTION 28.6 (Learning Outcome 27)

The arguments are those developed in Section 25.2.1, and summarized in point 5 of Section 25.4. The Group II peroxides decompose on heating as follows:

$$MO_2(s) = MO(s) + \tfrac{1}{2}O_2(g) \tag{Q.33}$$

The oxide ion, O^{2-}, in the decomposition product, MO, is smaller than the peroxide ion, O_2^{2-}. Thus, decomposition occurs to give a product with a larger value of $-L$, and stability increases with cation size. BaO_2 will therefore be the most stable of the peroxides, and is most likely to be formed on heating the Group II metals in oxygen.

QUESTION 28.7 (*Learning Outcome 27*)

The decomposition reaction is

$$MCO_3(s) = MO(s) + CO_2(g) \tag{28.26}$$

In Figure Q.4, a thermodynamic cycle has been built around this reaction. This gives:

$$\Delta H_m^{\ominus} = -L(MCO_3, s) + L(MO, s) + C$$

$$= \frac{8W}{r(M^{2+}) + r(CO_3^{2-})} - \frac{8W}{r(M^{2+}) + r(O^{2-})} + C$$

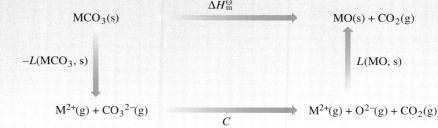

Figure Q.4 A thermodynamic cycle for discussing the stabilities of Group II carbonates.

As $r(CO_3^{2-})$ is obviously greater than $r(O^{2-})$, a particular increase in $r(M^{2+})$ will induce a larger decrease in $8W/[r(M^{2+}) + r(O^{2-})]$ than in $8W/[r(M^{2+}) + r(CO_3^{2-})]$. Thus, $\Delta H_m^{\ominus}$ will become more positive as the cation size increases, and stability will increase down the Group. Figure 25.7 covers the essentials of the arguments by using a different example.

QUESTION 28.8 (*Learning Outcome 26*)

In Section 28.1, we related the rates of reaction of the Group II metals in water to the solubility of their protective oxide/hydroxide films in the solvent. In water, both beryllium and magnesium are quite well protected, but because BeO is amphoteric and MgO is not, the protective film on beryllium is soluble in alkali, and the exposed metal then reacts quickly with water giving hydrogen gas. With the other Group II metals, the product would be a solid or dissolved hydroxide, $M(OH)_2$, but the amphoteric nature of $Be(OH)_2$ leads to formation of $[Be(OH)_4]^{2-}(aq)$:

$$Be(OH)_2(s) + 2OH^-(aq) = [Be(OH)_4]^{2-}(aq) \tag{Q.34}$$

Thus the overall reaction is,

$$Be(s) + 2OH^-(aq) + 2H_2O(l) = [Be(OH)_4]^{2-}(aq) + H_2(g) \tag{Q.35}$$

QUESTION 28.9 (*Learning Outcomes 26*)

Two kinds of explanation are possible. Firstly, as electronegativity increases across a Period, the electronegativity of magnesium will be greater than that of sodium, and the electronegativity difference between magnesium and chlorine will be less than that between sodium and chlorine. Secondly, Mg^{2+} is both smaller and more highly charged than Na^+, so it is more capable of polarizing the chloride ion. In either explanation, the implication is that $MgCl_2$ is less ionic than $NaCl$, and this is reflected in the structure.

QUESTION 28.10 (Learning Outcomes 26 and 28)

The scale appears because equilibrium in dissolution reactions such as

$$CaCO_3(s) \rightleftharpoons Ca^{2+}(aq) + CO_3^{2-}(aq) \qquad (Q.36)$$

lies well to the left. When $[edta]^{4-}(aq)$ is added, it removes $Ca^{2+}(aq)$ by forming a complex with these ions. The equilibrium tries to minimize the constraint by supplying more $Ca^{2+}(aq)$. The scale must dissolve to achieve this.

QUESTION 28.11 (Learning Outcomes 26 and 28)

According to Table 26.1, ethylenediamine acts as a bidentate ligand through the non-bonded pairs on its two nitrogen atoms. As beryllium seems to prefer tetrahedral coordination in its complexes, each beryllium will be tetrahedrally bound to two bidentate ethylenediamine ligands in a complex $[Be(en)_2]^{2+}$ (**Q.1**). The chloride will therefore be $[Be(en)_2]Cl_2$.

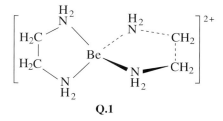

Q.1

FURTHER READING

1   M. Mortimer and P. G. Taylor (eds), *Chemical Kinetics and Mechanism*, The Open University and the Royal Society of Chemistry (2002).

2   L. E. Smart and J. M. F. G. Gagan (eds), *The Third Dimension*, The Open University and the Royal Society of Chemistry (2002).

3   C. J. Harding, R. Janes and D. A. Johnson (eds), *Elements of the p Block*, The Open University and the Royal Society of Chemistry (2002).

ACKNOWLEDGEMENTS

Grateful acknowledgement is made to the following sources for permission to reproduce material in this book:

Figures

Figure 1.2, 2.1: © The British Museum; *Figure 2.2*: Britain on View; *Figures 2.4 and 16.2*a: Martyn F. Chillmaid/Science Photo Library; *Figure 2.8*b: Delta Gold; *Figure 3.2b*: R. Maden, National Geographical Society; *Figures 3.6a and 12.3a* : Royal Society of Chemistry; *Figure 3.6b*: Fabre Minerals specimen, F. Fabre; *Figure 3.8*: Ben Johnson/Science Photo Library; *Figure 3.12*: photo by Peter Garside, Alcoa World Alumina, Australia; *Figure 3.14a*: photo supplied by Central Japan Railway Company with cooperation of the Japan Aluminium Association; *Figure 3.14b*: Mark Wagner/Flight Collection; *Figure 6.4*: Supplied by www.polytechphotos.com; *Figure 7.1 and 23.4*: Science Photo Library; *Figure 8.4*: Spencer Swagner/Tom Stack and Associates; *Figure 11.1*: from the Scientific Papers of J. W. Gibbs. Vol. 1, Longman Green (1906); *Figure 12.1b*: Photo RMN-Herve Lewandowski-Louvre; *Figure 12.3b*: US Department of Energy/Science Photo Library; *Figure 15.1a*: A. P. Photo; *Figure 15.1b*: Werner Burgess Fortean Picture Library; *Figures 17.1 and 28.12*: The British Museum; *Figure 17.6*: *Surviving the Iron Age* by Peter Firstbrook; *Figure 20.5*: Godfrey Argent Studio; *Figure 20.6*: F. Fabre; *Figure 21.2*: Tek Image/Science Photo Library; *Figure 23.2*: courtesy of the Wieliszka Salt Mines, Poland; *Figure 25.5*: V. Yatsina/Novosti; *Figure 28.2*: Hershel Freidman; *Figure 28.3*: Britain on View; *Figure 28.5*: Canadian Space Agency/ Agence Spatiale Canadienne 1996. Received by the Canada Centre for Remote Sensing, processed and distributed by RADARSAT International; *Figure 28.15*: Jerry Schad/Science Photo Library.

Every effort has been made to trace all the copyright owners, but if any has been inadvertently overlooked, the publishers will be pleased to make the necessary arrangements at the first opportunity.

Case Study

Batteries and fuel cells

Ronald Dell and David Johnson

INTRODUCTION

1

Before discussing batteries, we shall revisit electrochemical cells in general, and introduce some new terminology. Electrochemical cells are of two basic types: *voltaic* or **self-driving cells**, and *electrolytic* or **driven cells**. Self-driving cells were discussed in Section 21 of the main text. In a self-driving cell, chemicals react and generate an electrical current that can be used as a source of electrical power. However, in driven cells, an external source of current is passed through the cell to effect electrolysis and to generate chemicals. In industry, for example, driven cells are used to produce aluminium (Section 3.3), and chlorine and sodium hydroxide as twin products in sodium chloride electrolyis (Section 24.1).

Figure 1.1 is a simplified picture of a zinc–chlorine self-driving cell. One electrode consists of metallic zinc deposited on graphite; the other is graphite in contact with chlorine gas. Both electrodes make contact with an acid solution of zinc chloride, $ZnCl_2$, containing the ions $Zn^{2+}(aq)$ and $Cl^-(aq)$.

The reaction between zinc and chlorine is thermodynamically favourable. In Figure 1.1, it occurs as two separate electrode reactions. Zinc at the left-hand electrode is oxidized:

left-hand electrode: $Zn(s) = Zn^{2+}(aq) + 2e^-$ (1.1)

Chlorine gas at the right-hand electrode is reduced:

right-hand electrode: $Cl_2(g) + 2e^- = 2Cl^-(aq)$ (1.2)

If the two electrode reactions are added, the electrons cancel, and the overall cell reaction is obtained; it is the formation of aqueous zinc chloride solution from zinc and chlorine:

$Zn(s) + Cl_2(g) = Zn^{2+}(aq) + 2Cl^-(aq)$ (1.3)

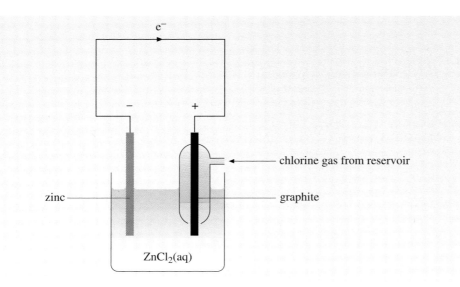

Figure 1.1
A self-driving cell operating in discharge mode: a spontaneous cell reaction produces electric current in an external circuit that bridges the two electrodes. Here, the cell reaction involves the combination of zinc and chlorine.

In Figure 1.1, the left-hand electrode has a negative charge because each dissolving zinc atom leaves two negatively charged electrons on the electrode. Likewise, each dissolving chlorine molecule removes two electrons from the right-hand electrode, so the latter is positively charged. The accumulation of like negative charges at the left-hand electrode is dispersed by the movement of electrons from left to right in the external circuit. The flow of current continues until much of the zinc and chlorine has been consumed: the cell has been in discharge mode and is now discharged. But all is not lost. We can now try to start the whole process again by regenerating zinc and chlorine through electrolysis.

Figure 1.2 shows how the self-driving cell of Figure 1.1 can be turned into a driven cell. A source of electrical power must be incorporated into the external circuit to drive electrons in the right-to-left direction. In Figure 1.2, this is a powerful battery; the long vertical stroke represents the positive electrode, and the short stroke represents the negative electrode.

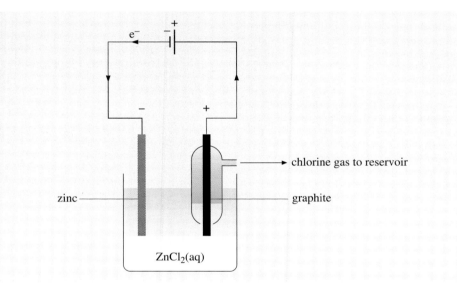

Figure 1.2 In a driven cell, a reaction that is not spontaneous is made to occur by the expenditure of energy in a power source inserted into the external circuit. Now the current flow and the reaction are reversed relative to Figure 1.1: zinc and chlorine are produced from aqueous zinc chloride.

Notice that the battery in the external circuit drives the (negative) electrons towards the left-hand electrode. This therefore accumulates negative charge, and is marked negative in Figure 1.2. Likewise, the associated withdrawal of electrons from the right-hand electrode leaves this with a positive charge. Thus, in both Figure 1.1 and Figure 1.2, each type of electrode has the same charge: zinc is negative and the chlorine electrode is positive.

The cell is now undergoing electrolysis: the Zn^{2+}(aq) ions are reduced to zinc metal at the left-hand electrode:

left-hand electrode: $Zn^{2+}(aq) + 2e^- = Zn(s)$ (1.4)

The Cl^-(aq) ions are oxidized to molecules of chlorine gas at the right-hand electrode:

right-hand electrode: $2Cl^-(aq) = Cl_2(g) + 2e^-$ (1.5)

The overall cell reaction, the sum of Reactions 1.4 and 1.5 is now:

$$Zn^{2+}(aq) + 2Cl^-(aq) = Zn(s) + Cl_2(g) \qquad (1.6)$$

By itself, this reaction (the reverse of Equation 1.3) is thermodynamically unfavourable. But it has been made to occur by coupling it to the favourable reaction that takes place in the battery in the external circuit: overall, the *combination* of the two reactions is thermodynamically favourable. The result is that our cell has been recharged with zinc and chlorine: it has been through a charging mode, and is now ready to be used again as a self-driving cell to generate electrical power.

In reading about batteries, you will often see electrodes described as anodes or cathodes. Whether the cell is in charge or discharge mode, the **anode** is the electrode where oxidation takes place, and the **cathode** is the site of reduction.

● In Figures 1.1 and 1.2, identify the anode and cathode.

● In Figure 1.1, oxidation occurs on the left, zinc electrode, and reduction at the right, chlorine electrode. Thus, in Figure 1.1, zinc is the anode and chlorine is at the cathode. In Figure 1.2, reduction occurs at the left and oxidation at the right. Zinc is now the cathode, and chlorine is at the anode.

We have already noted that in both the self-driving (discharge) and driven (charge) modes, the sign of a particular electrode is the same. Thus, in both Figures 1.1 and 1.2, zinc is negative. So, in switching modes, the sign of a particular electrode is unchanged, but the cathode and anode change places.

BATTERIES

2

A battery consists of a self-driving cell or cells. The zinc–chlorine system that we looked at in Section 1 is potentially an example of a **secondary battery**: it can be recharged many times, and for this reason has been explored as a possible power plant for electric vehicles. But most batteries that you use are not like this. They are **primary batteries**, which are discharged once and then thrown away.

The most important attribute of a battery is *portability*, allowing electrical devices to be operated away from a mains supply of electricity. A battery has certain essential components. The negative electrode consists of a metallic current collector or 'terminal' and an active component, which takes part in the electrode reaction. The latter is often a finely divided metal such as zinc, lead, cadmium or lithium, which can be oxidized with the release of electrons. The positive electrode also consists of a metallic current collector and an active component, generally a metal oxide (e.g. MnO_2, PbO_2, $NiO(OH)$, AgO), in which the metallic element is in a highly oxidized state. For example, if $NiO(OH)$ is thought of as a combination of O^{2-}, OH^- and Ni^{3+}, then nickel is in oxidation state +3. In a cell, $NiO(OH)$ is often reduced to $Ni(OH)_2$, in which nickel has its more common, lower oxidation state of +2. The electrodes are separated by an electrolyte, which conducts ions, but which must be an electronic insulator to avoid internal short circuits. In most conventional batteries the electrolyte is an aqueous solution such as $ZnCl_2$, KOH or H_2SO_4, although some advanced batteries use organic electrolytes, ion-conducting ceramics, polymers or molten salts.

The positive and negative electrodes of a battery cell are generally placed close together to minimize the internal resistance of the cell (which leads to loss of voltage when drawing a current). To prevent the electrodes touching inadvertently, so causing an internal short circuit, a mechanical **separator** is employed, and this is another essential cell component. Often it is in the form of a microporous sheet of polymer to hold the electrodes apart, but which absorbs electrolyte and permits ions to pass. These separators make little contribution to the cell resistance and mass.

For some applications, such as electric cars, a multi-cell battery consisting of a network of identical cells may be required. In this network, some cells may be connected *in series* as in Figure 2.1a; this increases the voltage. Others may be connected *in parallel* (Figure 2.1b.); here, the voltage of the combination is only that of a single cell, but the time for which a particular level of current can be delivered is much increased. It is important that the voltages of parallel cells remain identical after installation. Since the cell voltage changes somewhat during a cell's lifetime, it is advisable that parallel cells should be of comparable age. To see why, consider any two of the three cells in Figure 2.1b. They are connected in the same sense as are the battery and driven cell of Figure 1.2. If one has a higher voltage than the other, it will discharge, and charge the cell of lower voltage, as in Figure 1.2. So, if parallel cells are not well matched in voltage (as a result of a variation in their ages), the cells of lower voltage will drain current from their neighbours. Thus, in a large multi-cell battery, like the emergency power supply in a building, it is not uncommon for internal currents to circulate through parallel cells even when no external load is connected.

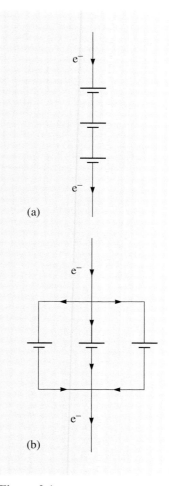

Figure 2.1
Cells connected (a) in series; and (b) in parallel.

BATTERY APPLICATIONS AND SIZES

3

Over the past 50 years the applications for small sealed batteries in the home have expanded phenomenally. Today, small primary or rechargeable batteries are employed in a huge number of appliances. Some examples (Figure 3.1) are as follows.

- *Household*: telephones, clock-radios, security alarms, smoke detectors, portable fluorescent lamps, torches and lanterns, door-chimes, car central-locking activators.

- *Workshop and garden*: portable tools (e.g. screwdrivers, drills, sanders), portable test meters, hedge trimmers, lawnmowers.

- *Entertainment*: portable radios and televisions, compact disc players, tape recorders, remote controllers for televisions and videos, electronic games and toys, keyboards.

- *Personal hygiene and health*: toothbrushes, bathroom scales, hair trimmers, shavers, blood-pressure monitors, hearing aids, heart pacemakers.

- *Portable electronic devices*: watches and clocks, cameras, camcorders, calculators, organizers, mobile phones, laptop and notebook computers, bar-code readers.

Figure 3.1
A range of batteries of different shapes and sizes.

Many of these applications require batteries of advanced design that give greatly improved performance as a result of developments in materials science and technology. Although most small consumer batteries are still of the primary ('throw-away') variety, there is a growing trend to adopt rechargeable batteries because they are more economical.

Moving to larger secondary batteries, principally lead–acid, there has been a growth in the market for engine-starter batteries to mirror the growth in transport. Every internal combustion engine, whether it be in a car, a truck, a bus, an aircraft or ship, requires a starter battery. These were once 'starting, lighting, ignition (SLI) batteries', reflecting their principal functions. With vehicles becoming more sophisticated, the need for electric motors and other electrical facilities has mushroomed. The demands on the battery have grown correspondingly, and it may soon be necessary to fit two batteries to the private car; one for engine starting, the other for running auxiliaries. At the same time, 'SLI batteries' have become 'automotive batteries' to reflect the other functions of the modern car battery.

There has been a similar rise in demand for very large, installed battery packs. Almost every public building (airports, hospitals, hotels, railway stations, stores and supermarkets, etc.) must have an uninterruptible power supply (UPS), so a battery pack can take over seamlessly when the mains supply fails, until such time as a local generator can be started. Other applications are shown in Table 3.1. In the past decade, new rechargeable batteries have been introduced into civilian and consumer markets. Moreover, there are now specialized batteries with unconventional chemistries for use in the military and the space fields.

As Table 3.1 shows, the size range of batteries is now enormous. The energy content of a battery is expressed in watt hours (W h). As $1\,W = 1\,J\,s^{-1}$, and $1\,h = 3\,600\,s$, $1\,W\,h = 3\,600\,J$. The scale extends from small button cells of energy content ~0.1 W h, all the way to load-levelling batteries of energy content ~10 MW h. Few other commodities come in such a range of sizes. Some typical applications for batteries of differing size are also given in Table 3.1.

Table 3.1 Battery sizes and applications

Battery type	Stored energy/W h	Applications
miniature/button cells	0.1–0.5	watches, calculators
portable communications	2–100	mobile phones, laptops
domestic uses	2–100	portable radio and TV, torches, toys, videocameras, power tools
automotive	10^2–10^3	starter batteries for cars, trucks, buses, boats, etc.; traction for golf carts, invalid chairs, etc.
remote area power supplies	10^3–10^5	lighting, water-pumping, telecommunications
traction	10^4–10^6	electric vehicles, forklift trucks, tractors
stationary	10^4–10^6	standby batteries
submarine	10^6–10^7	underwater propulsion
load levelling	10^7–10^8	electricity supply: load levelling, spinning reserve

CELL DISCHARGE AND CHARGE

As the current that a battery delivers increases, its voltage tends to fall. To understand this, we return to the cell considered in Section 21 of the main text. The method described there for determining a voltage for this cell is shown in Figure 4.1. An adjustable counter-voltage is inserted into the external circuit. When the counter-voltage, V, *just* stops the current generated by the cell, its value is equal to the cell voltage, E. The cell voltage, E, obtained at the point of balance is often called the *electromotive force*, or e.m.f. of the cell.

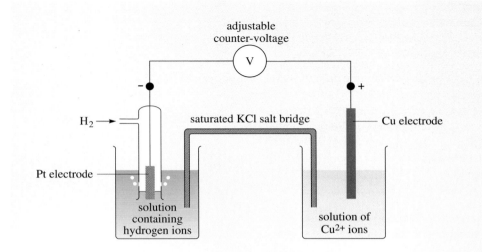

Figure 4.1 Measuring the voltage or e.m.f. of a reversible electrochemical cell. When the counter-voltage V in the external circuit is less than E, the e.m.f. of the cell, hydrogen gas reduces copper ions, and electrons flow from left to right in the external circuit. When V is greater than E, electrons flow from right to left, copper is oxidized, and hydrogen ions are reduced. Such a cell is said to be **reversible**. When E and V are equal, there is no current and no reaction.

In Section 21 of the main text, we also said that the reversible, zero-current voltage, E, is the maximum possible voltage that the cell can deliver. When the cell is discharging, and delivering current to the wire, its discharge voltage, is less than E. This is serious. It is useful to think of the voltage as the energy supplied to the external circuit per passed electron. If the voltage drops, the energy per electron drops. Discharge of the cell therefore means loss of energy that might have been used to do useful work by, say, driving an electric motor placed in the external circuit. Why does current flow cause this?

Notice first that whatever the size of the current, the energy liberated by a given amount of reaction in the cell is the same. The difference is that when a substantial current is flowing, less of this energy is available to do work. If it is not expressed as work, it must appear as heat. So the drop in the voltage when a cell delivers a significant current is associated with a greater evolution of heat.

One such source of heat is the internal resistance of the cell. The current flowing in the external circuit must also pass through the cell. There, it encounters resistance, just like a current flowing through the resistance of the wire heating element in an electric fire. The result in both cases is evolution of heat. In the case of the cell, this is initiated within the electrolyte.

There are also heat losses at the cell electrodes. Suppose Figure 4.1 is brought to the zero current situation with a stopping voltage. Now we reduce the stopping voltage very slightly so that a tiny current flows in the wire. The current is so small that the polarity of the electrodes is not disturbed. Here, the discharge voltage, V_d, is effectively equal to E. Now we lower the counter-voltage by a much greater amount, so that a substantial current flows in the wire. The flow of electrons out of the negative (left-hand) electrode tends to reduce the density of the negative charge; the flow of electrons into the right-hand electrode tends to reduce the density of positive charge. These sudden losses cannot be entirely made good by adjustments that they prompt elsewhere in the cell. So the initiation of a significant steady current diminishes the polarity difference between the two electrodes. This effect, like that arising from the internal resistance of the cell, tends to make the discharge voltage, V_d, less than E. The greater this deficiency, the greater the evolved heat.

Such losses of voltage arise from changes of polarity. They are called **polarization losses**, and they can occur at both electrodes, and also in the electrolyte. Polarization losses arise because when a steady current is started, the response of the resulting chemical changes initiated within the cell is never fast enough to restore the zero current polarities. For example, in a typical electrode, such as the nickel oxide positive electrode employed in rechargeable alkaline-electrolyte batteries, the reversible discharge reaction may be shown as:

$$NiO(OH)(s) + H_2O(l) + e^- \rightleftharpoons Ni(OH)_2(s) + OH^-(aq) \qquad (4.1)$$

In the left-to-right reaction, Ni^{3+} is being reduced to Ni^{2+}. For this process to take place, an electron from the current-collector has to interact with a water molecule and a particle of the solid nickel oxy-hydroxide. On a macroscopic scale, this will begin with particles of solid immediately adjacent to the current-collector, since these are encountered first by the emerging electron. As discharge proceeds, a reaction front will propagate into the bulk of the active material and away from the current-collector. Most batteries perform better if allowed to 'rest' or 'recuperate' periodically during discharge. This permits diffusion to take place, which reduces concentration gradients in the solids taking part in the battery reaction, and restores a balanced situation, nearer to equilibrium. On charging the battery, the reverse processes take place, again with diffusion controlling the reaction paths.

The cell voltage, therefore, drops increasingly below the reversible, zero-current value as the current drawn from it is increased. Figure 4.2 shows how this and similar effects work in practice for the charge and discharge of a small sodium–metal-chloride cell. The horizontal axis, with the unit **ampere hours** (A h), shows the charge that remains available in the cell at different stages of the discharge/charge cycle. In SI units, charge is measured in coulombs (C), and $1\,A\,h \equiv 3\,600\,C$. The discharge voltages (lower blue curves) lie below the reversible value of 2.35 V. At 9 A h, after 1 A h has been drawn from the cell, the discharge voltages decrease from about 2.25 V when the current is 5 A, to about 1.51 V when the current is 20 A. Each discharge curve also shows that, if a constant current is drawn from the cell, the voltage falls as the charge still available in the cell decreases; that is, the cell voltage decreases with time.

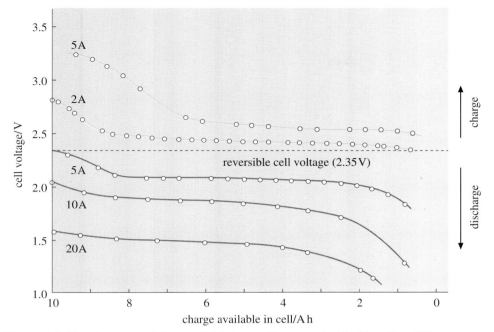

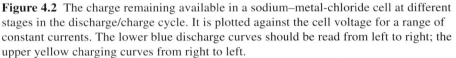

Figure 4.2 The charge remaining available in a sodium–metal-chloride cell at different stages in the discharge/charge cycle. It is plotted against the cell voltage for a range of constant currents. The lower blue discharge curves should be read from left to right; the upper yellow charging curves from right to left.

Notice that whereas the discharge voltages lie below the reversible value, the voltages for the charging curves lie above it. On discharge, the effects of polarization and internal resistance cause a voltage drop. During charging, such effects have to be overcome, and the required voltage *exceeds* the reversible value.

Ampere hours is also the unit used as a measure of the **storage capacity** of a cell or battery. This is the product of the number of hours for which the cell may be discharged at a constant current to a defined cut-off voltage at which the cell is no longer very useful. The value of the capacity depends not only on the ambient temperature and on the age/history of the cell, but also, to a greater or lesser extent, on the rate of discharge employed. The higher the rate of discharge, the less the available capacity. At the same time, as we have seen, the operating voltage falls off markedly at high discharge rates because of polarization. The result is that the **stored energy delivered**, measured in **watt hours** (W h = V × A h), declines even more sharply at higher discharge rates. When considering the capacity and the stored energy available from a battery, it is essential to define both the discharge rate and the temperature to be employed. Battery manufacturers generally state a **rated capacity** (so-called **nominal** or **name-plate capacity**) under specified discharge conditions, often the 5-h rate at 25 °C to a designated cut-off voltage. This can be determined by finding the constant current that causes a fall to the cut-off voltage after exactly 5 hours.

BATTERY SPECIFICATION

5

The specification for a battery depends very much on its proposed use. There are three broad considerations: technical performance, environmental factors (safety and recycling or disposal), and cost. Technical performance is affected by cell chemistry, materials of construction, design, ambient temperature, rate of discharge, depth of discharge before recharging, cut-off voltage on recharge, etc. We list below some of the desirable attributes looked for in a battery.

All batteries

- high cell voltage and stable voltage plateau over most of the discharge
- high stored energy content per unit mass ($W h kg^{-1}$) and per unit volume ($W h litre^{-1}$)
- low cell resistance (milliohms)
- high peak power output per unit mass ($W kg^{-1}$) and per unit volume ($W litre^{-1}$)
- high sustained power output
- wide temperature range of operation
- long inactive shelf life (years)
- long operational life
- low initial cost
- reliable in use
- sealed and leak proof
- rugged and resistant to abuse
- safe in use and under accident conditions
- made of readily available materials which are environmentally benign
- suitable for recycling

Secondary batteries

For these we may add:
- high electrical efficiency (W h output/W h input)
- capable of many charge/discharge cycles
- ability to accept fast recharge
- will withstand overcharge and overdischarge
- sealed and maintenance-free

This is a formidable list, and compromises have to be accepted. The relative importance of the various factors varies widely according to the application. For example, reliability under all conditions of use is relatively unimportant for an automotive battery; if it fails prematurely it is merely an inconvenience and a cost consideration. Contrast this with an aircraft battery, where sudden failure in flight would be much more serious, or with a battery powering a satellite, where — there being no access to the faulty item — failure would result in the loss of a multi-million pound vehicle.

A further complication is that the above desirable factors are often highly inter-active. For example, the available stored energy and the peak power output both depend on the temperature; the peak power also depends on the state of charge of the battery; the charge–discharge cycle life of a secondary battery depends critically on the severity of the discharge in each cycle, and so on. All these factors need to be quantified before we can decide whether a battery is likely to be commercially viable for a particular application.

5.1 Primary batteries

By far the most common primary cells are based on the zinc/manganese dioxide system, either so-called 'zinc–carbon cells' (Leclanché cells) or alkaline manganese dioxide cells. These both give 1.5 V open circuit, but differ in a number of important respects. Zinc–carbon cells (Figure 5.1a) have a central carbon rod immersed in the positive active material, or cathode (a mixture of compressed impure MnO_2 and carbon), a container of metallic zinc as the negative electrode, or anode, and an electrolyte of aqueous NH_4Cl and/or $ZnCl_2$ These cells are traditional and inexpensive. The chemistry is complex because there are several competing cell reactions. Here we give just one of them. At the anode, zinc undergoes its usual oxidation to Zn^{2+}; at the cathode, the oxidation state of manganese is reduced from +4 in MnO_2 to +3 in MnO(OH):

$$MnO_2(s) + H_2O(l) + e^- = MnO(OH)(s) + OH^-(aq) \tag{5.1}$$

As zinc hydroxide, $Zn(OH)_2$, is almost insoluble in water, the zinc and hydroxide ions combine to precipitate this solid, and the overall reaction is:

$$Zn(s) + 2MnO_2(s) + 2H_2O(l) = Zn(OH)_2(s) + 2MnO(OH)(s) \tag{5.2}$$

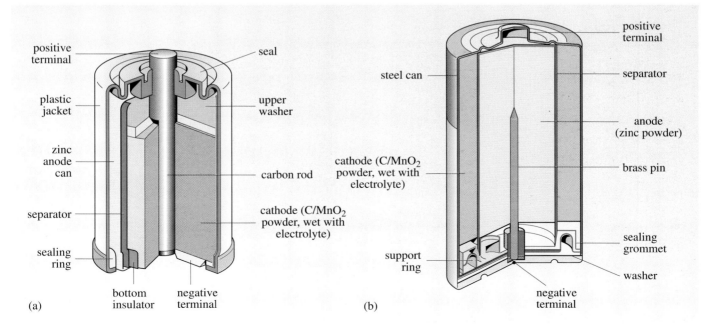

Figure 5.1 (a) Zinc–carbon battery; (b) alkaline manganese dioxide battery.

Alkaline manganese dioxide cells (Figure 5.1b), a superior and more expensive product, use finely divided zinc powder as the negative electrode, and this fills the centre of the cell, with a brass pin to make contact with the base. The electrolyte is concentrated KOH solution, and the positive electrode material, a mix of chemically or electrochemically prepared MnO_2 and carbon, forms a concentric annulus around the zinc powder and the separator. Alkaline manganese dioxide cells have a long shelf life and are particularly useful for high-drain (power) applications, where their useful life is several times that of zinc–carbon. The cheaper zinc–carbon cells are adequate for low-drain applications and for intermittent use (such as in torches), where there is recovery time between uses to allow diffusion processes to remove polarization at the electrodes and restore equilibrium. Both types of cell are made by most manufacturers in a variety of standard sizes and shapes. The prismatic (rectangular) 9 V batteries, as used in smoke detectors, contain six small cells connected in series.

Higher voltages can be obtained by using cells containing anodes made of metals whose oxidation is thermodynamically more favourable than that of zinc. Lithium (see Tables 12.1, 15.1 and 21.2 of the main text) is one such candidate, and several manufacturers are now offering 3 V lithium–MnO_2 cells. Water cannot be used in the electrolyte because lithium reacts with it, so these cells employ a lithium foil negative electrode and an ion-conducting organic electrolyte. The latter consists of a lithium salt, such as $LiAsF_6$ or $LiCF_3SO_3$, dissolved in a polar organic solvent such as 1,2-dimethoxyethane (Structure **5.1**), or tetrahydrofuran (Structure **5.2**). They are available as cylindrical cells, using spirally wound electrodes (so-called 'jelly roll' configuration, Figure 5.2a), or as miniature (button or coin) cells. In spirally wound cells, the lithium anode in the form of sheet foil is laid flat; a separator sheet is superimposed on it, and finally there is a sheet of bonded MnO_2 serving as the cathode. The three layers are laminated by passing through rollers under pressure

$$H_3COCH_2CH_2OCH_3$$

dimethoxyethane

5.1

tetrahydrofuran

5.2

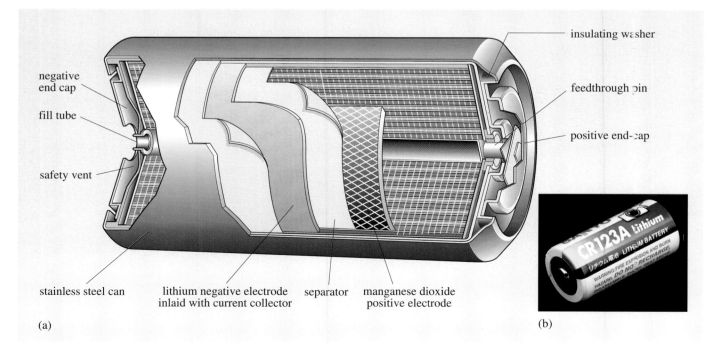

Figure 5.2 (a) High-rate lithium/manganese dioxide primary cell, spirally wound; (b) the commercial product.

and then cut to size and rolled into a cylindrical 'Swiss roll', which just fits in the can. Because the component layers are thin, and the rolling is tight, there is a far greater surface area of electrode in this type of cell than in a conventional alkaline–manganese cell, resulting in lower internal resistance. This is needed because the ionic conductivity of the organic electrolyte is much lower than that of concentrated KOH solution. As lithium is such a powerful reducing agent, with a density only half that of water, lithium batteries deliver relatively high energies per unit mass. They also have a long shelf life and the ability to operate over a wide temperature range ($-40\,^\circ$C to $+60\,^\circ$C).

Lithium button and coin cells are widely used in watches, clocks and pocket calculators. Among the possible cathodes are MnO_2 or a polymer with an empirical formula close to CF. This can be produced by the action of fluorine on graphite. In CF (Figure 5.3), the graphite rings have become puckered, each carbon forms one C—F bond, and the multiple bonding in the rings has been eliminated. Fluorination to a composition CF_x, where x is 0.5–0.95 leaves enough multiple bonding in the rings to provide some conducting properties, and the polymer can then be used as a cathode. If we write it as $(CF_{0.8})_n$, the cathode reaction is:

$$C_nF_{0.8n}(s) + e^- = C_nF_{0.8n-1}(s) + F^-(solv) \tag{5.3}$$

But lithium fluoride is insoluble in the organic solvent, so when fluorine enters it as fluoride ion, it combines with $Li^+(solv)$ to give solid lithium fluoride. The overall cell reaction is therefore:

$$Li(s) + C_nF_{0.8n}(s) = LiF(s) + C_nF_{0.8n-1}(s) \tag{5.4}$$

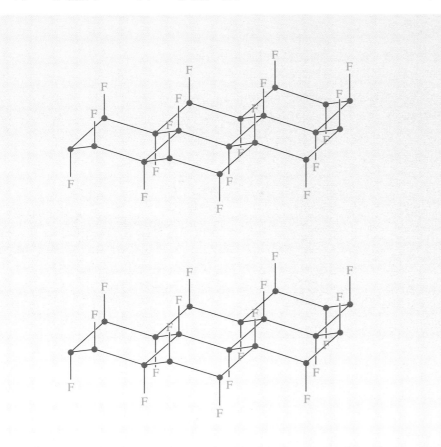

Figure 5.3
Structure of the layers of graphite after fluorination to the composition $(CF)_n$. When fluorinated graphite is used as a battery cathode, the degree of fluorination is reduced.

Alternative systems of lower voltage include zinc–manganese alkaline (1.5 V) and zinc–silver oxide cells (1.5 V). Altogether, there are over 40 different sizes and chemistries of button and coin cells.

In the military field, large batteries have been used to propel torpedoes. The MU-90 torpedo (Figure 5.4), developed for the French and Italian navies, relies on an aluminium–silver oxide cell with an alkaline electrolyte. $Al(OH)_3$ is amphoteric, and in alkali dissolves to form $[Al(OH)_4]^-(aq)$ (Section 3.3 of the main text). The silver oxide cathode consists of AgO in which silver is in oxidation state +2. The cell reaction is:

$$2Al(s) + 2OH^-(aq) + 3H_2O(l) + 3AgO(s) = 2[Al(OH)_4]^-(aq) + 3Ag(s) \quad (5.5)$$

The guidance systems of missiles are powered by large, lithium primary batteries, which use a molten lithium salt as electrolyte. The battery operates at >300 °C, and is heated rapidly to operating temperature by a pyrotechnic charge.

5.2 Secondary batteries

5.2.1 Lead–acid batteries

By far the most widely used rechargeable battery is the lead–acid battery (Figure 5.5). This finds applications in engine starting (automotive batteries), supplying domestic electrical loads for caravans and boats (leisure batteries), stand-by power supplies for hotels, shops, hospitals, factories, etc. (stationary batteries), for powering electric vehicles (traction batteries), and as back-up for solar and wind-powered installations. It differs from most other rechargeable batteries in that the electrolyte (sulfuric acid) participates in the electrode reactions, and is more than just an ionic conductor. This is made clear in the following electrode reactions:

negative electrode

$$Pb(s) + H_2SO_4(aq) = PbSO_4(s) + 2H^+(aq) + 2e^-; E^\ominus = 0.356 \text{ V} \quad (5.6)$$

positive electrode

$$PbO_2(s) + H_2SO_4(aq) + 2H^+(aq) + 2e^- = PbSO_4(s) + 2H_2O(l); E^\ominus = 1.685 \text{ V} \quad (5.7)$$

overall cell reaction:

$$Pb(s) + PbO_2(s) + 2H_2SO_4(aq) = 2PbSO_4(s) + 2H_2O(l); E^\ominus = 2.041 \text{ V} \quad (5.8)$$

During discharge, sulfuric acid is consumed and water is formed, with the converse on charging. The density of the electrolyte varies according to the density of sulfuric acid, which allows the density of the electrolyte to be used as a measure of the state of charge of the battery.

Although the lead–acid battery dates back to the nineteenth century, many improvements in its design and construction have been made, and, indeed, are still being made. Many rely on the use of new materials for the 'inactive' components (current collectors, separators, seals, battery cases, etc.) and on new manufacturing technology. Consequently, modern lead–acid batteries are greatly superior in performance to those of fifty years ago, and, in real terms, they are cheaper. Their principal limitation still lies in their excessive weight, a consequence of the high density of the solid lead. This places a fundamental limitation on the specific energy ($W h kg^{-1}$) that may be achieved with a lead–acid battery. Most practical lead–acid batteries deliver just 30–40 $W h kg^{-1}$.

Figure 5.4
The battery-powered MU-90 European torpedo.

Figure 5.5
The familiar lead/acid starting battery found beneath nearly every car bonnet.

5.2.2 Alkaline batteries

Another well-known rechargeable battery dating back 100 years is the nickel–cadmium battery. These are commonly found in small rechargeable appliances such as toys, portable cassette players and toothbrushes (Figure 5.6), as well as in much larger sizes for use in aircraft and as traction batteries for electric vehicles. The large ones are superior to lead–acid batteries in many respects, in particular their long life of up to 2 000 charge–discharge cycles, but they are also much more expensive.

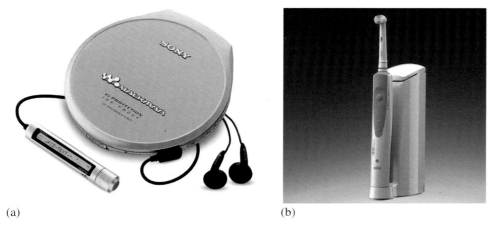

(a)   (b)

Figure 5.6   The CD player, and the electric toothbrush are two common devices often powered by rechargeable nickel–cadmium batteries.

The nickel–cadmium battery is an example of a class of rechargeable battery with an alkaline (KOH) electrolyte. The electrode reactions are:

negative electrode

$$Cd(s) + 2OH^-(aq) = Cd(OH)_2(s) + 2e^-; E^\ominus = 0.81 \text{ V} \tag{5.9}$$

positive electrode

$$2NiOOH(s) + 2H_2O(l) + 2e^- = 2Ni(OH)_2(s) + 2OH^-(aq); E^\ominus = 0.49 \text{ V} \tag{5.10}$$

overall reaction

$$Cd(s) + 2NiOOH(s) + 2H_2O(l) = Cd(OH)_2(s) + 2Ni(OH)_2(s); E^\ominus = 1.30 \text{ V} \tag{5.11}$$

At 1.30 V, the voltage of this battery is only two-thirds that of the lead–acid battery, and cadmium is also a relatively dense metal. As a result of these two factors, the specific energy of the nickel–cadmium battery is little better than that of the lead–acid cell. However, the nickel–cadmium type has other more marked advantages. Besides the longer cycle life already noted, it has a flatter discharge voltage, better low temperature performance, a continuous overcharge capability without causing damage, excellent reliability and low maintenance requirements.

In recent years a new alkaline nickel oxide battery has been developed and is now widely used in, for example, mobile phones (Figure 5.7). This is the so-called 'nickel–metal hydride' battery, usually abbreviated as Ni/MH. This has the same nickel oxide positive electrode and the same KOH electrolyte as a nickel–cadmium battery, but a new negative electrode in the form of a metal hydride. The negative electrode reaction is

$$MH_x(s) + OH^-(aq) = MH_{x-1}(s) + H_2O(l) + e^- \tag{5.12}$$

and the positive electrode reaction is the same as for the nickel–cadmium battery. The overall cell reaction is then:

$$NiO(OH)(s) + MH_x(s) = Ni(OH)_2(s) + MH_{x-1}(s) \qquad (5.13)$$

The practical cell voltage lies in the range 1.2–1.3 V, which is almost the same as nickel–cadmium batteries, and the two batteries are interchangeable. The metal hydride in this battery, which has been specially developed and optimized for it, is formed from a complex alloy of several metals, based either on lanthanide elements and nickel, or on titanium–zirconium. The specific energy of Ni/MH batteries is 60–70 W h kg^{-1}, almost twice that of Ni–Cd. Also, they are resilient to overcharge and over-discharge, have a high specific power output, and may be used over a wide temperature range (−30 to +45 °C). Given the toxicity of cadmium, which poses an environmental risk when disposing of used batteries, it seems likely that metal hydride batteries will progressively take over from cadmium batteries for many applications.

5.2.3 Lithium batteries

The desirable qualities of lithium batteries were mentioned in Section 5.1. Among the metals, it is the least dense and one of the most reactive. These qualities make it an obvious candidate for battery use, providing a much higher electrochemical reduction potential than zinc; its batteries have specific energies of some 130 W h kg^{-1}. The problems in developing lithium batteries stem from the high reactivity of lithium metal, particularly with water. This necessitates using a dry room for handling the metal and constructing cells. It is also necessary to use a non-aqueous electrolyte, which may be either an organic liquid or a solid polymer, each with a dissolved lithium salt to make it ionically conducting.

As already mentioned, *primary* lithium batteries using a lithium foil negative electrode, an organic liquid electrolyte, and any one of several positive electrode materials are commercially available. The positive electrodes used include MnO_2, MoS_2, V_6O_{13} and a 'carbon fluoride' CF_x (Figure 5.3). The difficulties arise when one tries to develop a *rechargeable* lithium battery of this type. This involves the electroplating of lithium metal from an organic electrolyte. Much work has been done in this field, with only limited success. Lithium is not normally electrodeposited as a smooth layer on the metal current collector, but as a 'mossy' deposit. Lithium foil that has been exposed to air is covered with a thin layer of hydroxide and nitride, which offers some measure of protection from further corrosion. Freshly electrodeposited lithium is finely divided and highly reactive, and decomposes the electrolyte. This is highly detrimental, and can be dangerous if it leads to build-up of gas pressure within the cell. Some of the metal deposit becomes electrically isolated from the electrode, and so capacity is lost rapidly. Lithium metal may also plate out in the form of tree-like crystals known as 'dendrites'. These can ultimately penetrate the separator, bridge the electrodes, and cause an internal short-circuit of the cell. Cells have been known to ignite spontaneously during recharge. For all these reasons, the commercial prospects for rechargeable cells based on liquid electrolytes and lithium metal negatives are not too promising, although much research is still in progress, particularly using solid polymer electrolytes. It is the lithium-ion battery, the rising star of recent years, which circumvents most of these problems.

The essential feature of the lithium-ion battery is that at no stage in the charge–discharge cycle should any lithium metal be present. Rather, lithium is incorporated

Figure 5.7
Many mobile phones use rechargeable nickel–metal hydride batteries.

('intercalated') into the positive electrode in the discharged state and into the negative electrode in the charged state, and it moves from one to the other through the electrolyte as Li^+ ions. The electrolyte is a solution of a lithium salt in an organic solvent. Figure 5.8 shows a schematic diagram of a secondary battery of this sort.

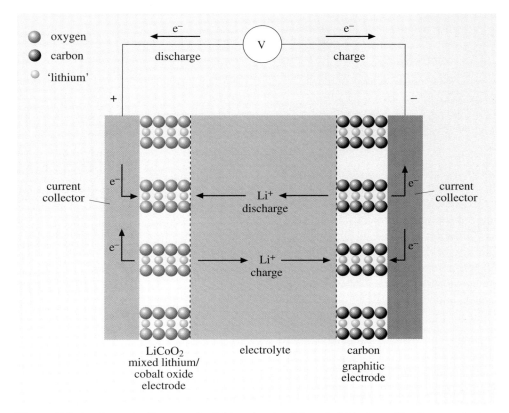

Figure 5.8 A schematic diagram of a lithium ion 'rocking-chair' battery. The item marked 'lithium' in the key represents the sites between which lithium ions move during charge and discharge. On discharge, lithium moves out from between the layers of the carbon atoms in the graphitic electrode as Li^+ ions. These pass through an electrolyte consisting of a lithium salt dissolved in an organic solvent or polymer. They then enter an oxide of a transition metal such as cobalt at the other electrode, and become bound to oxygen. In charging mode, lithium ions move back from the oxide to the graphite.

The origin of the cell voltage is the difference in the Gibbs function of lithium in the crystal structures of the two electrode materials. Commercial cells use carbon as the negative electrode because the Gibbs function change for the reaction of lithium metal with graphite is relatively small. The lithium in this material therefore retains most of its thermodynamic strength as a reducing agent. Lithium will intercalate readily into graphite up to a composition approaching C_6Li. This material is a metallic alloy. If we use a simple electron-gas model of metallic bonding, one can envisage the intercalated lithium as Li^+ ions lying between the planes of carbon atoms.

As you know, lithium undergoes a very energetic reaction with oxygen to form the oxide, Li_2O (Table 15.1 of the main text), in which lithium ions are surrounded by oxide ions. So positive electrodes that can supply an oxide environment for lithium ions are very suitable for lithium batteries. Such electrodes are normally based on either cobalt oxide or nickel oxide. The cells are assembled in the *discharged* state using a positive electrode of $LiCoO_2$ or $LiNiO_2$ from which Li^+ ions are 'de-intercalated' on charging, thereby transferring them into the carbon negative electrode. A 3 V lithium-ion cell results.

After a few initial charge–discharge cycles, approximately one-half of the intercalated lithium may be removed reversibly from the positive electrode, as shown:

$$\text{charged} \qquad\qquad\qquad \text{discharged}$$

$$Li_{0.55}CoO_2 + 0.45Li^+ + 0.45e^- \rightleftharpoons LiCoO_2 \text{ (123 A h kg}^{-1}\text{)} \qquad (5.14)$$

$$Li_{0.35}NiO_2 + 0.5Li^+ + 0.5e^- \rightleftharpoons Li_{0.85}NiO_2 \text{ (135 A h kg}^{-1}\text{)} \qquad (5.15)$$

The values in brackets give the practical electrode capacity per kilogram of positive active material for the de-intercalation of the amounts of Li^+ shown.

Lithium-ion cells are constructed in the same 'spiral wound' configuration as primary lithium cells (Figure 5.2). When cycling Li^+ ion cells, it is important to control the top-of-charge voltage carefully (4.1 V for $LiNiO_2$ and 4.2 V for $LiCoO_2$). Failure to do so results in decomposition of the positive electrodes to give oxygen gas and Co_3O_4 or $LiNi_2O_4$, a hazardous situation in a sealed cell. For this reason, lithium-ion cells must be recharged using a specially designed charger incorporating both voltage and temperature control. Over-discharge must also be avoided, and it is usual to have a limiting cut-off voltage on discharge of ca 2.7 V. (In this regard the Ni/MH battery has the advantage of being much better able to withstand overcharge and overdischarge).

Much research is in progress on the possibility of using a lithium–manganese oxide as the positive active material in place of $LiCoO_2$ and $LiNiO_2$. This would lead to cost reductions, since manganese is much cheaper than cobalt or nickel. Unfortunately, the structure of the lithium–manganese oxide system is less easy to control, and the development of a fully satisfactory lithium-manganese oxide electrode has not proved to be easy.

The SONY Corporation in Japan first commercialized lithium-ion cells in the early 1990s, and other battery manufacturers now also make them. They are widely used in laptop computers (Figure 5.9), mobile phones and other portable electronic equipment. In 1998 alone, about 280 million cells were manufactured in Japan. With much research and development in progress, we can expect future improvements in performance as well as price reductions.

Figure 5.9
The high specific energy of rechargeable lithium batteries accounts for their frequent use in laptop computers.

DEGRADATION MODES IN BATTERIES

6

If a battery is to be charged/discharged for hundreds, or even thousands, of cycles, it is essential that the chemical reactions taking place at the electrodes are quantitatively reversible. Even if as little as 0.1% irreversibility (or side-reaction) occurs, this will soon add cumulatively to a major loss in capacity. Many, if not most, electrode reactions involve a reconstructive phase change in the crystal chemistry of the active materials. A typical positive electrode reaction would be

$$\text{solid A} + \text{anion} \rightleftharpoons \text{solid B} + e^- \tag{6.1}$$

$$(\text{e.g. Ni(OH)}_2\text{(s)} + \text{OH}^-\text{(aq)} \rightleftharpoons \text{NiOOH(s)} + \text{H}_2\text{O(l)} + e^-)$$

This involves ionic diffusion processes in the crystal structure of the solids, leading to phase change and recrystallization. The need to reverse this reaction *quantitatively* during each cycle is a very demanding one. The severity of the specification is apparent when one considers the many possible processes or side-reactions leading to battery deterioration and failure. These include:

- growth of metallic needles at the negative electrode, causing internal short circuits;

- mechanical shedding of active material from electrode plates;

- separator dry-out through over-heating;

- corrosion of current collectors, resulting in increased internal resistance;

- gas formation at electrode plates on overcharge, causing disruptive effects.

These and other degradation processes may result in sudden battery failure, through an internal short circuit, or may lead to progressive loss in capacity and performance. Generally, the degradation steps are interactive and accumulative, so that, when the performance starts to deteriorate, it soon accelerates and the battery becomes unusable. Nevertheless, remarkable success has been achieved in designing batteries of long cycle life (~1 000 cycles) for several different chemistries. Our modern telecommunications, weather forecasts and military defences are totally dependent on the performance of such batteries in satellites, where lifetimes of > 20 000 cycles are sometimes realized. Figure 6.1 shows one of the more recent examples, for which a nickel–metal hydride battery has been chosen.

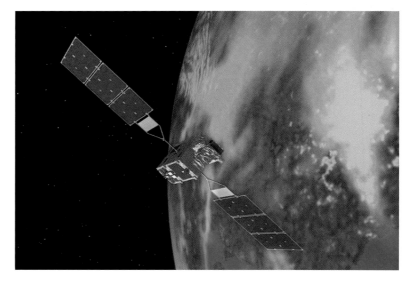

Figure 6.1
The Optical Inter-orbit Communications Test-Satellite (OICETS) is a joint European and Japanese venture. Its nickel–metal hydride batteries are recharged from the solar panels, here shown fully deployed.

FUEL CELLS

7

Fuel cells are essentially water electrolysers working in reverse; a fuel gas (normally hydrogen) is fed to one electrode and oxygen (or air) to the other. Electrochemical reaction takes place to form water, generating a voltage across the cell, with the fuel electrode being negative and the oxygen electrode positive. A direct current may then be drawn. A schematic diagram of a fuel cell is shown in Figure 7.1.

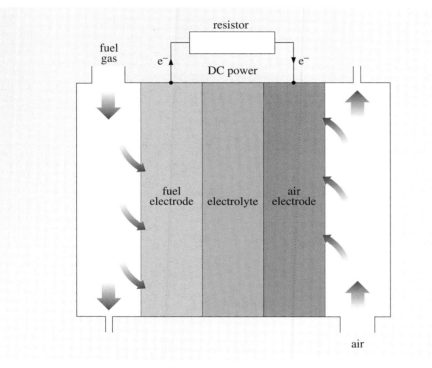

Figure 7.1 Schematic diagram of a fuel cell.

The fuel cell was invented in 1839 by Sir William Grove, and has been much investigated ever since on account of its perceived advantages for power generation. Compared with the internal combustion engine, fuel cells have high thermodynamic efficiency, and good performance under low load conditions. Other advantages include low pollution emission, rapid response time, simplicity of mechanical engineering and modular factory construction (which facilitates the gradual build-up of large units). It has, however, proved difficult to turn fuel cells into commercially viable power devices, using natural gas, oil or coal as primary fuels. Such fuels must first be 'reformed' (i.e. converted) to hydrogen, and the difficulty of developing a satisfactory reformer is comparable to the difficulty of developing the fuel cell itself. There is also the problem of poisoning of the catalyst used in the fuel reformer, and of the electrocatalyst in the fuel cell, by impurities (especially sulfur) in the primary fuel. Altogether, a fuel cell that uses primary fuels poses complex engineering problems.

With pure, electrolytic hydrogen there is no need for a reformer and no impurities to worry about. The problem of developing a satisfactory fuel cell is therefore much simpler. Hydrogen–oxygen fuel cells have been used in manned space flight, the product water being drunk by the astronauts. Plans have also been developed for combined electrolysers and fuel cells for this application, as a means of storing solar electricity generated by photovoltaic cells. Thus, space technology is providing a lead for future terrestrial applications.

A fuel cell differs from a conventional battery in that the reactants are gaseous and stored outside the cell. Therefore the capacity of the device is limited only by the size of the fuel and oxidant supply. For this reason, fuel cells are rated by their *power* output (kW) rather than by their stored *energy* (kW h). There is interest in fuel cells for both static (e.g. combined heat and power generation for buildings) and mobile applications. A particular attraction of fuel cells for mobile applications — for example, powering electric vehicles — is that the positive reactant (air) does not have to be transported, and is free.

Unfortunately, this advantage is offset by the difficulty of conveying hydrogen. Compressed hydrogen in gas cylinders is both bulky and heavy, liquid hydrogen is impractical for most uses, and hydride storage beds are still in experimental stages. So a fuel-cell powered car would be likely to use liquid methanol as a fuel, with a reformer on board to decompose it to hydrogen and carbon dioxide. This concept is now in an advanced development stage, with several major automotive companies (Daimler–Benz, Ford, etc.) involved. For larger vehicles, buses and trucks, it is possible that compressed hydrogen gas in cylinders might be used. Ultimately, the oil companies would favour the development of an on-board reformer for petroleum.

Fuel cells come in a number of different types, which differ in the electrolyte they employ and the temperature range over which they operate. The thermodynamic reversible voltage for the decomposition of water, calculated from the values of the Gibbs function, decreases linearly from $E^{\ominus} = 1.229$ V at 298 K to 1.088 V at 473 K.

Thermodynamics therefore favours the operation of a water *electrolyser* at relatively high temperature. In contrast, a fuel cell, being the reverse electrochemical reaction, is favoured by operation at low temperature and high pressure, but kinetic factors (electrocatalysis, polarization) call for higher temperatures and a compromise must be made. At high temperatures, the overall energy efficiency can be maintained by waste heat recovery, even though the net voltage efficiency is lower. Typically, fuel cells generate only 0.7–0.8 V.

The five principal types of fuel cell are summarized in Table 7.1. These have been developed for different applications. Phosphoric acid fuel cells operating at 200 °C are now manufactured in large (MW) unit sizes to supply combined heat and power (CHP) to large building complexes. Alkaline fuel cells, used first in spacecraft, have been developed in multi-kW size for powering electric vehicles. A problem with using an alkaline electrolyte is that it absorbs CO_2 from the air, poisoning the electrolyte and reducing its conductivity. So the incoming air must be pre-scrubbed to remove CO_2. This problem is avoided in the polymer electrolyte membrane fuel cell (PEM) in which the membrane is acidic. You met such membranes in the discussion of the electrolytic chlor-alkali industry in Section 24.1.1 of the main text. A typical membrane consists of a perfluorosulfonic acid polymer containing SO_2OH groups. The $-SO_2OH$ group is strongly acidic, and so allows the passage of hydrogen ions. That this is essential can be seen from the reactions at the two

Table 7.1 Principal types of fuel cell

Fuel cell type	Electrolyte examples	Temperature range/°C	Electrocatalyst
alkaline	KOH	50–150	nickel positive; steel negative (or Pt on C)
phosphoric acid	H_3PO_4	200	Pt on C
polymer electrolyte membrane (PEM)	perfluorosulfonic acid polymer	100	Pt on C
molten carbonate fuel cell (MCFC)	Li_2CO_3	650	Li_2O/NiO positive; NiCr alloy negative
solid oxide fuel cell (SOFC)	ZrO_2–Y_2O_3	700–1 000	$(La, Sr)MnO_3$ positive; Ni/ZrO_2–Y_2O_3 negative

electrodes:

negative electrode

$$2H_2(g) = 4H^+(mem) + 4e^- \qquad (7.1)$$

positive electrode

$$O_2(g) + 4H^+(mem) + 4e^- = 2H_2O(l) \qquad (7.2)$$

overall reaction

$$2H_2(g) + O_2(g) = 2H_2O(l) \qquad (7.3)$$

The two electrodes consist of carbon containing pores or channels, which allow passage of the reacting gases. The catalyst that promotes the two electrode reactions is platinum, and it is applied to the carbon in regions where carbon, gas-filled pores and proton-exchange membrane electrolyte all meet (Figure 7.2).

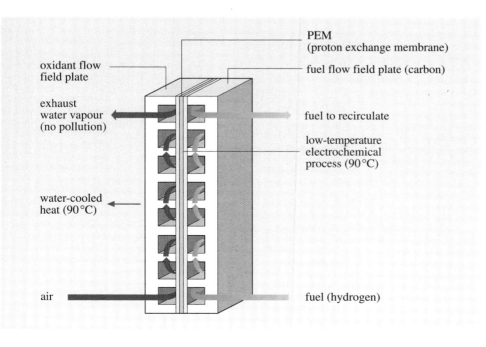

PEM
(proton exchange membrane)

oxidant flow field plate

fuel flow field plate (carbon)

exhaust water vapour (no pollution)

fuel to recirculate

low-temperature electrochemical process (90 °C)

water-cooled heat (90 °C)

air

fuel (hydrogen)

Figure 7.2

Diagram of a PEM fuel cell in which hydrogen is the fuel, and air is the source of oxygen.

The PEM fuel cell is one of the more promising types for powering electric vehicles, but the rather high cost of the perfluorosulfonic acid polymer membranes is a limiting factor. In North America, trials are proceeding with city electric buses powered by PEM fuel cell stacks. In Europe, Daimler–Benz have demonstrated their small A-class car in an electric version powered by a PEM fuel cell. This is the NECAR 3, where NE stands for 'no emissions' (Figure 7.3a). The pure hydrogen fuel is manufactured on-board by reforming methanol. In this case, reforming generates the greenhouse gas CO_2, so 'no emissions' is not strictly correct. However, this is not true of NECAR 4 (Figure 7.3b), in which the fuel is stored hydrogen. Rapid advances are being made in this technology and there is optimism that fuel-cell-powered cars will meet the requirements for a viable electric vehicle.

There are two high-temperature fuel cells, the molten carbonate fuel cell (MCFC) and the solid oxide fuel cell (SOFC). Both present difficult problems for materials science. Molten alkali carbonate at 650 °C is a most aggressive medium, and corrosion problems are severe in this fuel cell. The SOFC operates at even higher temperatures, in the range 700–1 000 °C, depending on the composition of the solid oxide electrolyte employed. The conduction mechanism is the movement of oxide ions (O^{2-}) through a defective oxide lattice. Here the problems are concerned with fabrication of ceramic shapes and the thermal expansion of components. There are also engineering problems relating to heat and mass transfer with both types of high-temperature fuel cell. Nevertheless, considerable research progress has been made, and prototypes of both the MCFC and the SOFC in the multi-kW range have been built and tested.

Fuel cell development requires the collaboration of physical chemists, materials scientists and chemical engineers. Good progress is now being made, and there is confidence that fuel cells of all five types can be manufactured that will operate well. When and where they will be introduced for terrestrial applications is now largely a matter of economics.

(a)

(b)

Figure 7.3
Two Daimler–Benz demonstrator cars powered by fuel cells. NECAR 3 (a) uses methanol as a fuel. It is converted to hydrogen in a catalytic reformer in the rear of the vehicle. In NECAR 4 (b), the fuel is liquid hydrogen, stored in an insulated tank in the same location.

ACKNOWLEDGEMENTS

Grateful acknowledgement is made to the following sources for permission to reproduce material in this book:

Figure 3.1: Colin Cuthert/Science Photo Library; *Figure 5.4*: courtesy of the Italian Navy; *Figure 5.5*: Charles D. Winters/Science Photo Library; *Figure 5.6a*: courtesy of Sony UK Limited; *Figures 5.6b and 5.9*: courtesy of Braun Oral Care; *Figure 6.1*: © NASDA; *Figure 7.3a and b*: courtesy of Ballard Power Systems.

Every effort has been made to trace all the copyright owners, but if any has been inadvertently overlooked, the publishers will be pleased to make the necessary arrangements at the first opportunity.

INDEX

Note Principal references are given in bold type; picture and table references are shown in italics.

CD-ROM INFORMATION

Computer specification

The CD-ROMs are designed for use on a PC running Windows 95, 98, ME or 2000. We recommend the following as the minimum hardware specification:

processor	Pentium 400 MHz or compatible
memory (RAM)	32 MB
hard disk free space	100 MB
video resolution	800 × 600 pixels at High Colour (16 bit)
CD-ROM speed	8 × CD-ROM
sound card and speakers	Windows compatible

Computers with higher specification components will provide a smoother presentation of the multimedia materials.

Installing the CD-ROMs

Software must be installed onto your computer before you can access the applications. Please run INSTALL.EXE from the CD-ROM.

This program may direct you to install other, third party, software applications. You will find the installation programs for these applications in the INSTALL folder on the CD-ROMs. To access all the software on these CD-ROMs, you must install QuickTime, Isis/Draw, WebLab ViewerLite and Acrobat Reader.

Running the applications on the CD-ROM

You can access *Metals and Chemical Change* CD-ROM applications through a CD-ROM Guide (Figure C.1), which is created as part of the installation process. You may open this from the **Start** menu, by selecting **Programs** followed by **The Molecular World**. The CD-ROM Guide has the same title as this book.

The *Data Book* is accessed directly from the **Start | Programs | The Molecular World** menu (Figure C.2), and is supplied as Adobe Acrobat document. If you are unfamiliar with Acrobat, please run the introduction supplied in the CD-ROM Guide.

Problem solving

The contents of this CD-ROM have been through many quality control checks at the Open University, and we do not anticipate that you will encounter difficulties in installing and running the software. However, a website will be maintained at

   http://the-molecular-world.open.ac.uk

which records solutions to any faults that are reported to us.

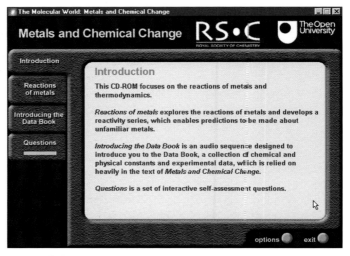

Figure C.1 The CD-ROM Guide.

Figure C.2 Accessing the *Data Book* and CD-ROM Guide.